To Leslie
Thanks!
Barry Shafer

The Ockham Brief

Cosmology, Quarks and High Energy Physics

The Ockham Brief

Cosmology, Quarks, and High Energy Physics

By

Barry Pierce Shafer

Sun Books
Sun Publishing Co.
Santa Fe, New Mexico

A ***Sun Books*** Publication

First Printing – 2013 June

Registration number: TXu 1-842-147
Email: Barry Pierce Shafer <rarllc02@earthlink.net>

Cover Illustration and Layout by **Barbara Elaine Brown**

Sun Books

Sun Publishing Company
P.O. Box 5588
Santa Fe, New Mexico 87502-5588 USA
info@sunbooks.com
www.ABookSource.com
www.SunBooks.com

10 digit ISBN: 0-89540-461-3
13 digit ISBN: 978-0-89540-461-9

The Ockham Brief

Cosmology, Quarks and High Energy Physics

Contents

I Introduction

I present a proposal for a new, coherent, and parsimonious scenario for Big-Bang cosmology [1]*. The present work is a combination, correction, and extension of two unpublished (1) (2) manuscripts "Quark Heuristics" [2] and "Gluon Heuristics at t = 0" [3]. Although I have used results derived from Quantum Electrodynamics and Quantum Chromodynamics [4] [5], my presentation is held to a fairly elementary level. In particular, the mathematical arguments involve only algebra and elementary differential calculus. This monograph should be readily comprehensible to a fairly wide audience.

Most of the chapters that follow have a common format. The chapters begin with a synopsis, followed by supporting analyses, endnotes, and references. In some cases, the chapters consist primarily of descriptive material, followed by endnotes and references.

In 1964, Murray Gell-Mann and George Zweig independently invented a theory of subatomic particles [6] that later became widely known as the Quark Model (3). Around the same time, Richard Feynman introduced an alternative concept known as the Parton Model [7]. A thorough understanding of the material presented in References [6] and [7] requires expert knowledge and abilities. Fortunately, a general introduction to the subject has been provided by Frank Wilczek, in his book The Lightness of Being [8]. On pages 43 and 44 of his book, Wilczek presents a fairly light-hearted description of the rivalry between the advocates of the quark and parton models. As Wilczek pointed out, both models had valid points. It turned out that the nucleons (neutrons and protons) that compose ordinary matter, are in turn composed of quarks and force-carrying particles called gluons (because they "glue"

---

* Parentheses ( ) refer to notes listed at the end of each chapter and brackets [ ] refer to references listed at the end of each chapter.

the quarks together to form nucleons).

The Particle Physics Booklet lists six quarks and their properties, such as mass, electric charge, and spin [9]. The six quarks are designated up, down, strange, charm, bottom, and top, {u, d, s, c, b, t}, and collected into three families or "generations", where {u, d}, {s, c}, and {b, t}, are the first, second, and third generations, respectively. The search for more generations is ongoing.

In the present work, I propose an extension of the presently accepted quark sequence {u, d, s, c, b, t} to a sequence comprised of 11 generations of quarks. The extension leads directly to a new, coherent, and parsimonious model for the Big-Bang, reasonable candidates for weakly-interacting massive particles (WIMPs), and a straightforward scenario for the unification of the strong, weak, electromagnetic and gravitational forces.

The extended quark sequence is postulated to result from a pair-production scenario, which I have termed the Ockham* process, in which a primordial gluon ensemble undergoes a transformation (or "condensation") into heavy quarks and antiquarks, resulting in both stable and unstable quark complexes. I postulate that the unstable quark complexes decayed into the extended quark sequence, and the stable quark complexes formed the WIMPs, that may compose the Dark Matter (DM) considered to be a major component of our Universe. I also discuss possible reasons for the predominance of "left-handed" electrons observed following beta decay.

---

* Named in honor of William of Ockham (1285 - 1347), credited with originating the principle of "Ockham's Razor". The spellings "Ockham" and "Occam" are often used interchangeably. In this monograph, the simpler spelling "Occam" is used more frequently.

Notes

(1) Earlier versions of [2] were submitted to Physical Review Letters (PRL) for publication, but were not accepted. I received the following from an Assistant Editor at PRL:

"PRL does not consider manuscripts which propose speculative alternatives to accepted theories unless certain conditions are met, conditions which in our judgement your manuscript did not meet."

I submitted the following rebuttal, but PRL held to their decision not to publish my manuscript (note, in particular, the conclusion in boldface):

"[Editor's name], what 'accepted theories' can you cite of the unification of the strong, weak, electromagnetic, and gravitational forces? None such 'accepted theories' exist! Consider the following from p. 3 of my manuscript:

'The discussion in [5]* implies two possibilities for the unification scale of the electroweak and gravitational forces, i.e. 10^{15} GeV or the Planck mass 1.2209×10^{19} GeV . The present work proposes the intermediate value of 6.9625×10^{17} GeV for the unification scale.'

However, [5] is from Chris Quigg's well known book on Gauge Theories, where we find the following on p. 294:

'Gravitation is omitted, although the unification scale is only four orders of magnitude from the Planck mass.'

---

* [5] refers to Reference [5] in [2], not Reference [5] at the end of this chapter.

Clearly, if my conjecture of a unification scale of 6.9625 x 10^{17} GeV holds, this would be quite significant and suggests the unification of the four fundamental forces."

Since this exchange with PRL, I have continued to develop my theories and I feel my case is even stronger. Although my proposed unification scale has changed slightly, the basic concepts have remained unchanged.

(2) Five distinguished physicists, at the Los Alamos National Laboratory, have reviewed and commented on these manuscripts. These individuals are listed in the acknowledgements at the end of Chapter XII.

(3) One version of the naming process (it may be apocryphal), has it that Zweig originally called his entities "aces", but Gell-Mann's catchier term "quarks" won out.

References

[1] K. A. Olive and J. A. Peacock, *Big-Bang Cosmology, Particle Physics Booklet,* U. S. Department of Energy, Pub, 72, July, 2004, p. 217.

[2] B. P. Shafer, *Quark Heuristics,* Unpublished Manuscript, Seventh Revision, March 10, 2008.

[3] B. P. Shafer, *Gluon Heuristics at t=0,* Unpublished Manuscript, January 16, 2010.

[4] A. Pich, *Quantum Chromodynamics,* Lectures given at the 1994 European School of High-Energy Physics (Sorrento, Italy, 29 August – 11 September, 1994), arXiv:hep-ph/9505231 v1 4 May 95.

[5] C. Quigg, *Gauge Theories of the Strong, Weak, and Electromagnetic Interactions,* Westview Press, 1997.

[6] Quigg, p. 24.

[7] Quigg, pp. 23, 164.

[8] F. Wilczek, *The Lightness of Being - Mass, Ether, and the Unification of Forces,* Basic Books, 2008.

[9] Particle Data Group, *Particle Physics Booklet,* U. S. Department of Energy, Pub, 72, July, 2004, pp. 25, 26.

II From Now to Then and Back Again

Synopsis

This chapter describes my initial thoughts on the subject of quark condensation and the proposal to extend the quark sequence. Later developments and improvements are discussed in the chapters that follow. The chapter title, "From Now to Then and Back Again", follows from my proposal to extend the quark sequence from the present ("Now"), backwards in time to Big-Bang time ("Then"), 13.7 ± 0.2 Gyr before present [1]. The proposed quark sequence extension is then assumed to provide a model for the evolution of the Universe from "Then" (Big-Bang time) back to the present ("and Back Again"). Hence, "From Now to Then and Back Again".

Analysis

About seven years ago (early in 2005) I noticed that the six quark masses, recorded in [2], appeared to follow a distinctive pattern, in particular, a linear-log relation between quark number and quark mass (m_q), where q = u, d, s, c, b, t, is the quark identifier. The present work was motivated by a linear-log plot of the mass ratios (m_q /m_u), for the first six quarks vs quark sequence number, minus one (Fig. II .1). A linear fit to all six quark masses (u, d, s, c, b, t) produced the lower line, with a slope of 2.1761. A linear fit to the s, c, and t masses produced the upper line, with a slope of 2.3939. The linear fits were clearly not perfect, but the linear-log correlations were good enough to motivate an attempt to extend a linear correlation to the ultra-high masses (on the order of the Planck mass) that likely appeared during the first instants of the creation of the universe.

The description of the radiation content of the early universe [3] presents a table that includes a schematic of the postulated production of

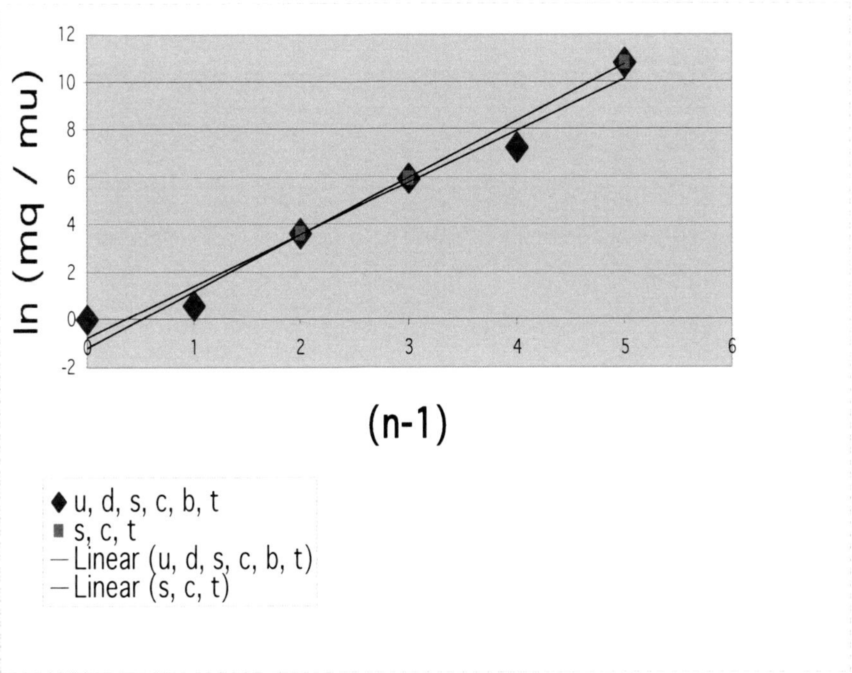

Fig. II.1
Ln (m_q / m_u) vs (n -1)
(quark masses m_q and m_u , n = quark sequence number)

quark pairs that accompanied the earliest moments of the expansion of the universe. I proposed that the pair production scenario be pushed back to the very beginning of the current expansion epoch. Assuming that the current epoch began with a spherical ensemble of gluons, consisting of N_0 gluon octets G_0, the outer radius of the sphere began moving into the void at velocity c_0 . I postulated that quark condensation began almost immediately, as the sphere expanded and cooled.

The next assumption was crucial. The first quark condensation was assumed to have occurred when the gravitational energy and the electrostatic energy of the topmost (most massive) quark pair were equal. This implies that quark mass and charge were "expressed" simultaneously. The assumption appeared to be reasonable because it resulted in an initial equipartition of energy between electrostatic and gravitational components. This condition was expressed by the equation $G_N m^2 = k Q^2$ (**II.1**), where G_N is the Newtonian gravitational constant, m is the quark mass, $k = 1/(4 \pi \varepsilon_0)$, ε_0 is the permittivity of free space, and Q is the topmost quark charge, assumed to be +2/3 e. Solving for m, we found $m = (k / G_N)^{1/2} Q = 6.9625 \times 10^{17}$ GeV (**II.2**), approximately 5.7% of the Planck mass of 1.2209×10^{19} GeV. The discussion in [4] implies two possibilities for the unification scale of the electroweak and gravitational forces, i.e. 10^{15} GeV or the Planck mass 1.2209×10^{19} GeV . As noted above, equality of the gravitational and electrostatic energies resulted in the intermediate value of 6.9625×10^{17} GeV for the unification scale. The topmost quark was denoted "mc+" (mass and +charge). Further work indicated that a second massive quark, with charge -1/3, should follow the mc+ quark. This quark was designated " mc-". Having 1/2 the charge of mc+, mc- is "expressed" at 1/4 the energy of mc+. In this work, both the electronvolt (eV) and eV/c^2, are commonly used as units of mass.

We considered some implications of the range of the strong force. Assume that the primordial gluon ensemble was composed of energy quanta having Planck masses and Planck dimensions. Assuming a value for the total mass of the universe (including dark matter and dark energy) of 3.00×10^{52} kg, we computed an initial radius for the universe of 1.116 fm, which is comparable to the size of a proton [5], and therefore comparable to the range of the strong force. Using similar reasoning, for the appearance of the mc+ quark, we obtained a radius of 19.6 fm, and for the appearance of the mc- quark, we obtained a radius of 78.3 fm, both values far in excess of the range of the strong force. Clearly, further expansion of the quark condensate would have been impeded little by the action of the strong force.

The next step was to extrapolate the line in linear-log space for the s, c, t quarks to the assumed topmost quark mass (6.9625 x 10^{17} GeV). The quark sequence number is the integer nearest the "mc+" intercept, plus one. For the extrapolations, two sets of quark mass estimates, consistent with [2], were used (Table II .1). For Set 1, the lower bound of 4.0 MeV was chosen for m_d and the simplistic assumption $m_d - m_u = m_n - m_p$ = (939.57 – 938.27) MeV = 1.3 MeV, was used to compute m_u . For Set 2, the quark mass ratios proposed by Weinberg and listed in [6], $m_u : m_d : m_s$ = 0.55 : 1 : 20.3, were used. Although Sets 1 and 2 produced nearly the same quark sequence number for the "mc+" quark, Set 2 was thought to be the better choice because the estimates of m_u and m_d are likely better. The extrapolation using Set 2 is presented in Fig. II.2. The extrapolation was obtained using an average of the linear fits presented in Fig. II .1 and results in the extrapolation presented in Fig. II.2. The nearest integer (n -1) to the topmost quark mass is 21, which means that n = 22. If the primordial left and right gluon octets are allowed to occupy positions 24 and 23, respectively, Table II.2 results, which supports the existence of 11 generations of quarks. The charge sequence is postulated to continue from the up quark (n = 1) to the topmost quark (n = 22), as follows: +2/3, -1/3, -1/3, +2/3, … ,+2/3, -1/3, +2/3.

Set	u	d	s	c	b	t
1	2.7	4.0	130	1,350	4,900	189,000
2	3.52	6.40	130	1,350	4,900	169,200

Table II .1
Quark Masses Used for Extrapolations (MeV)

At this point, it is useful to change the nomenclature and return to the conventional quark names, with modification, to accommodate the additional quark families. For convenience, we will use the usual quark

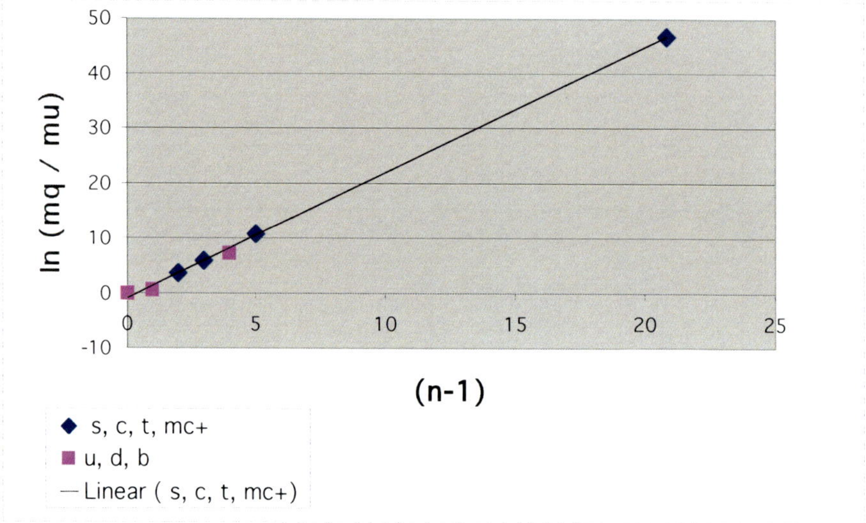

Fig. II .2
Extrapolation to Quark Mass "mc+"
Ln (m_q / m_u) vs (n -1)
(quark masses m_q and m_u , n = quark sequence number)

names, {u, d, s, c, b, t}, with {u, c, t} identified with +2/3 charge, and {d, s, b} identified with -1/3 charge. The indices 1, 2, 3, and 4, following the quark names, denote the quark families. Because we now have 3 complete quark families and 1 partial family, our extended quark sequence becomes {u1, d1, s1, c1, b1, t1, d2, u2, s2, c2, b2, t2, … , d4, u4, s4, c4}, as shown in Table II .2. Note that we have replaced mc+ and mc- by c4 and s4, respectively.

As noted at the beginning of this chapter, the primordial gluon ensemble was postulated to have expanded with the velocity of light and quark condensation was assumed to have begun at the outer surface of the gluon sphere and propagated inward. When the first quark and

14 u3 +2/3	16 c3 +2/3	18 t3 +2/3	20 u4 +2/3	22 c4 +2/3	G_{0L}
13 d3 - 1/3	15 s3 -1/3	17 b3 -1/3	19 d4 -1/3	21 s4 -1/3	G_{0R}
2 d1 -1/3	4 c1 +2/3	6 t1 +2/3	8 u2 +2/3	10 c2 +2/3	12 t2 +2/3
1 u1 +2/3	3 s1 -1/3	5 b1 -1/3	7 d2 -1/3	9 s2 -1/3	11 b2 -1/3

Table II .2
Extended Quark Sequence

antiquark, c4 and $\overline{c4}$ appeared, the gluon sphere became a quark-gluon plasma, and further quark condensation was assumed to have accompanied the expansion of the plasma sphere.

As the plasma sphere expanded, the average energy density decreased and more volume became available for quark condensation, and the appearance of the first quark eigenstates. As noted previously, the first quark and antiquark c4 and $\overline{c4}$ were expressed at a plasma sphere radius of 19.6 fm and the second quark and antiquark, s4 and $\overline{s4}$ were expressed at a plasma sphere radius of 78.3 fm.

Following the derivation of Fig. II .2 and Table II .2, further work was undertaken to derive the slope s of the extended quark sequence line from fundamental principles. The value of s found from Fig. II .2, was approximately 2.290, fairly close to ln (10) = 2.303. This appeared to be no mere coincidence. If s is actually ln (10), rather than 2.290, this implies a mass decrement of 10 between adjacent quarks, along the quark sequence line. In Chapter XI, we present proposals for the virtual processes that produce the mass decrement of 10 between adjacent quarks along the central portion of the quark sequence line.

As noted previously, at the top end of the quark sequence, the mass decrement between quarks c4 and s4, is 4. We proceed by assuming that the mass decrement between the remaining adjacent quarks in the extended quark sequence is 10. Finally, at the bottom end of the quark sequence, the mass decrement between quarks d and u is assumed to be 4. With these assumptions, a downward extrapolation from quark c4 to the u1 quark, results in Figs. II.3 and II .4. The agreement of the extrapolated quark masses with the masses from Set 2, Table II.1, is very encouraging.

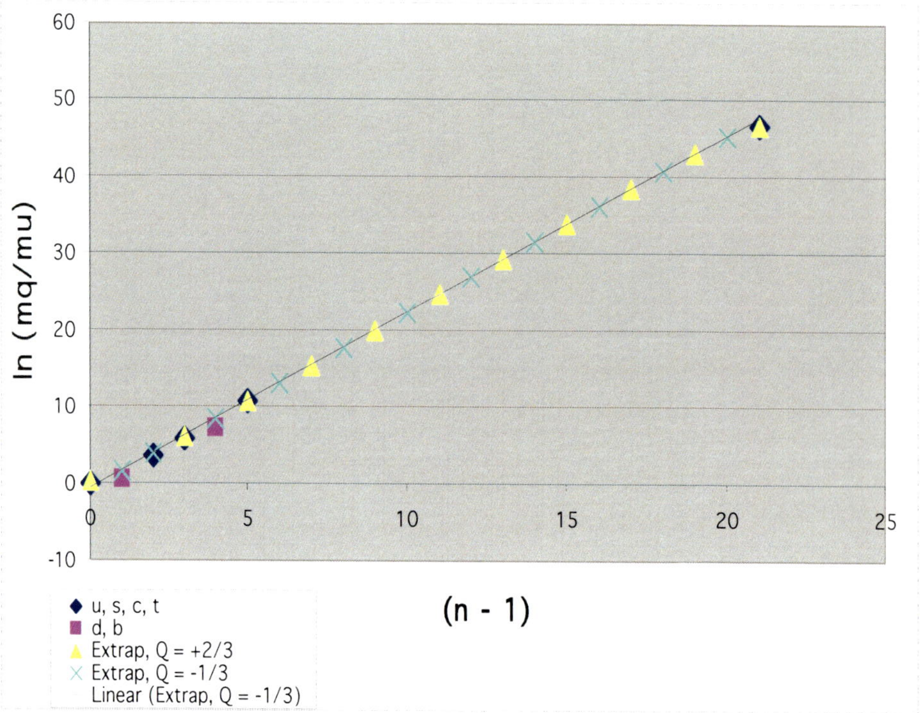

Fig. II .3
Downward Extrapolation from Quark Mass "mc+" or "c4"
Ln (m_q / m_u) vs (n -1)
(quark masses m_q and m_u , n = quark sequence number)

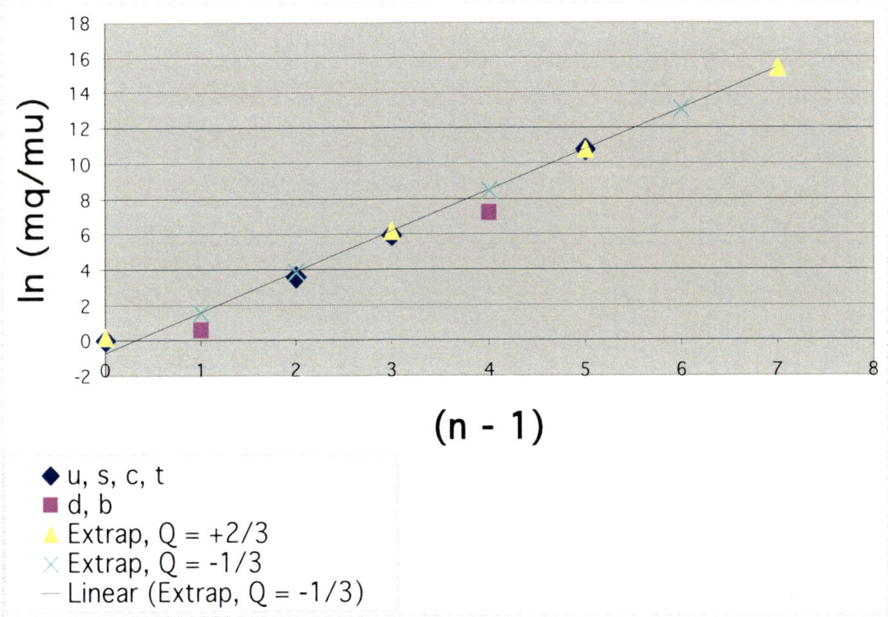

Fig. II .4
Downward Extrapolation from Quark Mass "mc+" or "c4" (detail at lower energies)
Ln (m_q / m_u) vs (n -1)
(quark masses m_q and m_u , n = quark sequence number)

However, the discovery of the Ockham Process led to slightly different and presumably more accurate values for the masses of the topmost quarks. The assumption $E_Q = E_G$ leads to the first approximation. The first unpublished manuscript *Quark Heuristics* [7] presented only this first approximation. An energy balance approach leads to a second, more accurate approximation.

Finally, as I stated at the beginning of this chapter, I began to consider an extended quark sequence early in 2005. It was only early in 2008 that I found a similar concept discussed in [8]. This reference also considered the possibility of an extended quark sequence, however, without discussing where the sequence might end, as I do in the present work.

References

[1] K. A. Olive and J. A. Peacock, *Big-Bang Cosmology, Particle Physics Booklet,* U. S. Department of Energy, Pub, 72, July, 2004, p. 221.

[2] Particle Data Group, *Particle Physics Booklet,* U. S. Department of Energy, Pub, 72, July, 2004, pp. 25, 26.

[3] Olive and Peacock, p. 223.

[4] C. Quigg, *Gauge Theories of the Strong, Weak, and Electromagnetic Interactions,* Westview Press, 1997, pp. 285, 294.

[5] T. P. Smith, *The Anatomy of a Neutron*, American Scientist, Vol. 98, No. 6, November – December, 2010, p. 478.

[6] A. Pich, *Quantum Chromodynamics,* Lectures given at the 1994 European School of High-Energy Physics (Sorrento, Italy, 29 August – 11 September, 1994), arXiv:hep-ph/9505231 v1 4 May 95, p. 44.

[7] B. P. Shafer, *Quark Heuristics,* Unpublished Manuscript, Seventh Revision, March 10, 2008.

[8] Quigg, p.152.

III The Lightness of Being

I obtained my copy of *The Lightness of Being* [1] in March of 2009. Written by Nobel Laureate Franck Wilczek, for general readers, the book presents an excellent overview of modern physics and cosmology. However, I discovered some minor errors (1), (2), and one process description in which I was in total disagreement. On page 92, Wilczek describes the appearance of the Universe from "nothing". On the previous page, he references a reservoir of "negative energy" stored in the vacuum. I disagree totally.

I believe energy is a positive scalar (a number > 0), period. Actually, this is not the first time I had encountered "negative energy". In the late 60s, I took a graduate physics course for which we used the text *The Atomic Nucleus* by Robley D. Evans [2]. On page 702, where Evans discusses pair production (referring to electron and positron "pairs") and the Dirac theory for pair production, Evans references "negative energy", which in my view, is really better treated as antimatter. For the remainder of this book, we shall assume that energy is a positive scalar. In the discussion that follows, all of the energetic processes will be assumed to involve positive quantities.

Notes

(1) On page 16, the mass of the "electron + positron" should be 2×10^{-27} gram, *not* 2×10^{-28} gram. This correction results in a relativistic multiplier of three thousand, *not* thirty thousand. Actually, this result is easily seen by "inspection" using mass units of MeV/ c^2 :

electron + positron : 1 MeV/ c^2
pion : 140 MeV/ c^2
10 pions : 1400 MeV/ c^2

proton, antiproton : 940 MeV/ c^2
proton + antiproton : 1880 MeV/ c^2
10 pions + proton + antiproton : 3280 MeV/ c^2
Multiplier : 3280 MeV/ c^2 / 1MeV/ c^2 = 3280

(2) On page 156, the Planck units should be obtained by using the reduced Planck constant h_r instead of h. This error results in a value of the Planck length too high by a factor of $\sqrt{(2\pi)} = 2.51$.

References

[1] F. Wilczek, *The Lightness of Being - Mass, Ether, and the Unification of Forces,* Basic Books, 2008.

[2] R. D. Evans, *The Atomic Nucleus*, McGraw-Hill Book Company, 1955, p. 702.

IV The Quantum Vacuum

Synopsis

A discussion of the quantum vacuum is essential to the understanding of the virtual processes that produce the weak decays of massive quarks, all the way down to the u and d quarks that compose the nucleons (neutrons and protons).

Analysis

The most common weak decay is the process d → u, which results in the beta decay of a neutron into a proton, electron, and electron-antineutrino, with a half-life of 886 s [1]. This process is shown in Fig. IV.1 . Note, in particular, the presence of a W^{-1} boson that appears as a "mediating particle" in the decay process.

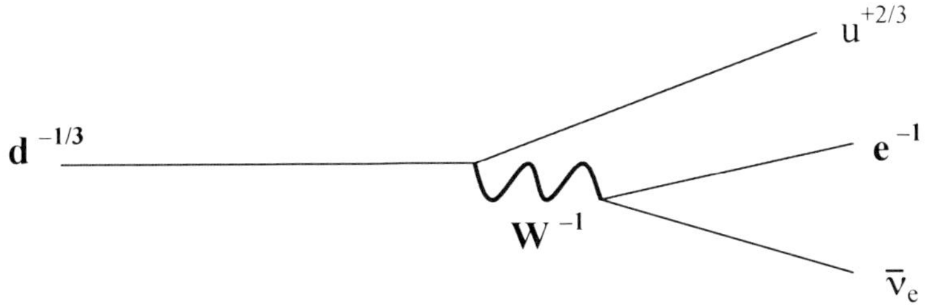

Fig. IV. 1
Beta decay, d → u, via mediating particle $\mathbf{W^{-1}}$.

From [2], we see that d has a mass of 3 – 7 MeV / c^2 and u has a mass of 1.5 – 3.0 MeV / c^2. However, W^{-1} is much more massive (80.4 GeV / c^2). Therefore, since the transformation d → u involves only a few MeV in energy, we well might ask, how can a W^{-1} boson, with energy 80.4 GeV, participate in the process $d^{-1/3} \rightarrow W^{-1} + u^{+2/3} \rightarrow u^{+2/3} + e^- + \bar{\nu}_e$?

The answer lies in the fact that W^{-1} is a "virtual" or very short-lived particle, that arises from and returns to the "quantum vacuum".

The uncertainty principle of quantum mechanics [3] specifies that a real particle having energy ΔE and lifetime Δt, must satisfy the condition $(\Delta E) \times (\Delta t) \geq h_r$, to have a physical existence,* where $h_r = h / 2\pi$, is the reduced Planck constant. If, on the other hand, we have the condition $(\Delta E) \times (\Delta t) < h_r$, the particle is virtual, arising as a brief fluctuation from the quantum vacuum. The lifetime of a particle of energy ΔE is limited by the uncertainty relation

$$\Delta t < h_r / \Delta E \qquad \text{(IV. 1)}.$$

Therefore, a whole host of particles can arise spontaneously from the quantum vacuum, as long as their lifetimes meet the condition (IV. 1).

In the discussion that follows, we represent gluon pairs ($g\,\bar{g}$) and quark pairs ($q\,\bar{q}$) as harmonic oscillators, with energy $E = h\nu = h_r\,\omega$. A quantum mechanical treatment of a harmonic oscillator is found in [4], where the quantized (discrete) energy levels are given by :

$$E_n = (n + 1/2)\, h_r \omega\,, \quad n = 0, 1, 2, \ldots. \qquad \text{(IV. 2)},$$

and the lowest, or ground state is given by :

$$E_0 = h_r\, \omega / 2, \qquad \text{(IV. 3)}.$$

As it turns out, Eq. IV.2 defines the boundary of the quantum vacuum in energy-frequency space, where the x-axis is the frequency (ω) and the y-axis is the energy (E). In action-frequency space, where the x-axis is frequency (ω) and the y-axis is action (A), the quantum vacuum boundary is defined by the relation:

$$A = h_r \qquad \text{(IV. 4)}.$$

---

* The energy ΔE is measured in Joules (J) and Δt is in seconds (s). The product $(\Delta E) \times (\Delta t)$ is called the "action" (A) and is measured in J s , as is h, Planck's constant.

Having defined the quantum vacuum boundaries (two straight lines in energy-frequency space and action-frequency space), we ask, is there an upper limit to frequency? In Chapter II, we discussed the extended quark sequence and the equality of the gravitational and electric dipole energies that occurs when the gluon energies decrease to about 5.7 % of the Planck energy and the postulated quark condensations that begin at these energies. It therefore seems reasonable that Planck units could be chosen as hypothetical upper limits for particle energies and hypothetical lower limits for particle sizes.

Combining Eq. IV.1 with the frequency of the harmonic oscillator, we can define the boundaries of the quantum vacuum in energy-frequency space and action-frequency space. In the energy-frequency space, the x-axis is the frequency (ω), and the y-axis is the energy (E). In the action-frequency space, the x-axis is the frequency (ω) and the y-axis is the action (A).

The vacuum is defined by :

$$\Delta t < h_r / \Delta E \quad \text{(IV. 1)},$$

and the vacuum boundary is defined by :

$$\Delta t = h_r / \Delta E \,.$$

For the oscillator (IV. 2),

$$\Delta E = E_n = (n + 1/2)\, h_r\, \omega \,,$$

and we take Δt as the quarter-period of the oscillator:

$$\Delta t = \tau / 4\,, \quad \text{where } \omega\, \tau = 2\pi\,,\ \tau = 2\pi / \omega\,,$$

and $\Delta t = \pi / 2\, \omega$.

Substitute into (IV.1) :

$$\Delta t < h_r / \Delta E$$

$$\pi / 2\, \omega < h_r / [(n + 1/2)\, h_r\, \omega\,]\,, \text{ then}$$

$$\pi / 2 < 1/ (n + 1/2)\,,\ n = 0,1,2, \ldots$$

Note that the inequality is satisfied only for n = 0:

$$\pi / 2 < 1/(0 + 1/2) = 2 \quad .$$

For n =1, we have :

$$1/(1 + 1/2) = 2/3,$$
$$\pi / 2 > 2/3,$$

and the inequality (IV.1) is violated. Therefore for the energy levels

$$E_n = (n + 1/2)\, h_r\, \omega ,$$

only E_0 lies within the quantum vacuum.

To find the boundary of the vacuum, solve the equation:

$$\Delta t = h_r / \Delta E$$
$$\pi / 2 = 1/(n + 1/2)$$
$$n + 1/2 = 2 / \pi$$
$$n = 2 / \pi - 1/2 = 0.137 ,$$

not an integer, but we can use this fraction to define the vacuum boundary. Therefore, we clearly see that E_0 lies within the vacuum, E_n ($n \geq 1$) lies above the vacuum, and $E_n = (2 / \pi)\, h_r\, \omega$ defines the vacuum boundary. For the action of the oscillator, we have :

$A = (\Delta E) \times (\Delta t) = [(n + 1/2)\, h_r\, \omega]\,(\pi / 2\omega) = (\pi / 2)(n + 1/2)\, h_r$

In action-frequency space, the vacuum boundary is given by (see Eq. IV.4) :

$$A = (\Delta E) \times (\Delta t) = h_r \, .$$

Comparing the oscillator action to the vacuum boundary, we have for $n = 0$, $A = (\pi / 2)(0 + 1/2)\, h_r = (\pi / 4)\, h_r < h_r$, and n = 0 is below the vacuum boundary.

For $n = 1$, $A = (\pi / 2)(1 + 1/2)\, h_r = (3\pi / 4)\, h_r > h_r$, and n = 1 is above the vacuum boundary.

Oscillator energies and actions for n = 0, 1, 2, ... are shown in Fig. IV. 2a and IV. 2b, along with the vacuum boundaries in the energy-frequency and action-frequency spaces. The vertical lines at $\omega = \omega_p$ represent the Planck frequency, defined by $\omega_p = E_p / h_r$, where E_p is the Planck energy, 1.2209×10^{19} GeV, the approximate energy at which the four fundamental forces ; strong, weak, electromagnetic, and gravitation are thought to unify.

In the present work, it is postulated that the heaviest quarks "condense" at a frequency somewhat below ω_p, whereas above ω_p, the only particles postulated to exist are gluons, with zero rest mass and zero electrical charge.

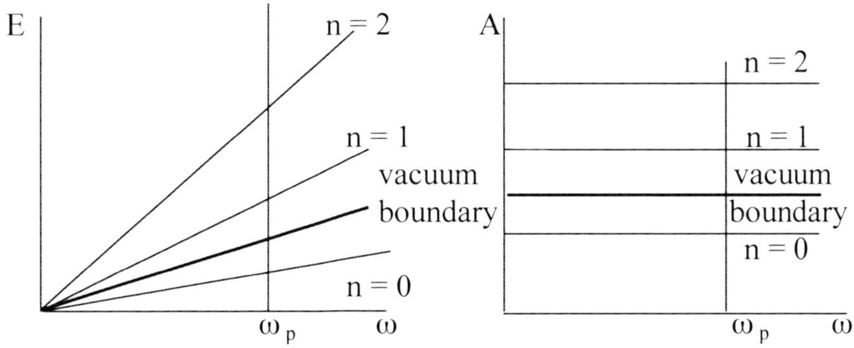

Fig. IV. 2a
Energy–Frequency Space

Fig. IV. 2b
Action–Frequency Space

As we discussed above, there does appear to be an upper limit to quark energies and harmonic oscillator frequencies that define these energies through the relations

$$E_n = (n + 1/2)\, h_r\, \omega_p \;, \; n = 0,1,2, \ldots$$

However, for the gluons that cool to eventually form the heavy quark sequence, there is no a priori reason to set an upper limit on gluon frequency, as long as frequency remains finite. In other words, initial gluon frequencies could be any finite number of multiples of the Planck frequency (ω_p).

Finally, I distinguish between the "vacuum" and the "void". I call the x-axes of both figures ($E = 0$ and $A = 0$), the void, and the regions between the lines ($E = 2\, h_r\ \omega / \pi$, $A = h_r$) , the vacuum. The void is totally empty and therefore manifestly stable.

References

[1] Particle Data Group, *Particle Physics Booklet,* U. S. Department of Energy, Pub, 72, July, 2004, p. 119.

[2] Particle Data Group, *Particle Physics Booklet,* U. S. Department of Energy, LBL-61033, July, 2006, p. 24.

[3] A. Messiah, *Quantum Mechanics, Volume I* , John Wiley and Sons, Inc., 1958, p. 135.

[4] A. Messiah, *Quantum Mechanics, Volume II* , John Wiley and Sons, Inc., 1958, p. 963.

V Initial Conditions

The description of the radiation content of the early universe [1] presents a table that includes a schematic of the postulated production of quark pairs that accompanied the earliest moments of the expansion of the universe. I proposed that the pair production scenario be pushed back to the very beginning of the current expansion epoch. I assume that the current epoch began with a mono-energetic spherical ensemble of gluons, consisting of N_0 gluon octets G_0 (divided equally between left and right-handed octets G_{0L} and G_{0R}). The initial gluon energy (E_0) is postulated to lie well above the Planck energy (E_P), i.e. $E_0 >> E_P$. The nine gluons depicted schematically as "dumbbells" in Fig. V.1, are shown in [2] to comprise the following gluon octet:

$g_1 = R\bar{B}$ $\qquad$ $g_5 = G\bar{R}$
$g_2 = R\bar{G}$ $\qquad$ $g_6 = G\bar{B}$
$g_3 = B\bar{R}$ $\qquad$ $g_7 = (2R\bar{R} - B\bar{B} - G\bar{G}) / \sqrt{(6)}$
$g_4 = B\bar{G}$ $\qquad$ $g_8 = (B\bar{B} - G\bar{G}) / \sqrt{(2)}$

where the first six gluons are singlets, and the seventh and eighth gluons are triplets and doublets, respectively. However, in the discussion that follows, we will refer to $R\bar{R}$, $B\bar{B}$, and $G\bar{G}$ as gluons also, even though they appear in gluons g_7 and g_8 as combinations (triplets and doublets) rather than singlets. R, B, G, $\bar{R}$, $\bar{B}$, and $\bar{G}$ denote red, blue, green, antired, antiblue, and antigreen, respectively.

The next chapters discuss, in detail, the Ockham process and evolution of the initial gluon ensemble into Regular Matter (RM) and Dark Matter (DM) by means of two distinct Ockham processes. We redefine the heaviest quarks c4 and s4, from Table II.2, as $U^{+2/3}$ and $D^{-1/3}$, respectively. Then, we postulate that the heaviest quarks $U^{+2/3}$ and $D^{-1/3}$ originate with the g_7 and g_8 gluons by means of Ockham processes, with the $U^{+2/3}$ quarks deriving from the $2R\bar{R}$ component of g_7 and the $D^{-1/3}$ quarks deriving from the $B\bar{B}$ and $G\bar{G}$ components of g_7 and g_8 .

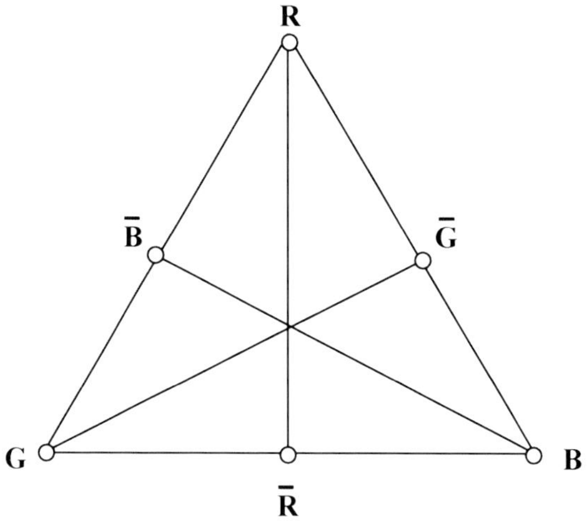

Fig. V.1
Nine components: $R\bar{R}$, $B\bar{B}$, $G\bar{G}$, $R\bar{G}$, $R\bar{B}$, $B\bar{R}$, $B\bar{G}$, $G\bar{B}$, and $G\bar{R}$, represented as "dumbbells" in the figure above, compose the gluon octet: g_1, g_2, g_3, g_4, g_5, g_6, g_7 and g_8.

References

[1] K. A. Olive and J. A. Peacock, *Big-Bang Cosmology, Particle Physics Booklet,* U. S. Department of Energy, Pub, 72, July, 2004, p. 223.

[2] C. H. Llewellyn Smith, *Quantum Flavordynamics, Quantum Chromodynamics, and Unified Field Theories*, edited by K. T. Mahanthappa and J. R. Randa, Plenum, New York, 1980, p. 61.

VI The Ockham Process – Part 1

The discussion of Big-Bang cosmology included a listing of new particles that appeared as the Universe cooled [1]. Most of the particles produced in the cooling process appear in pairs, e.g. $q,\bar{q}$, $W^{\pm}$, $l^{\pm}$ (quarks, bosons, and leptons). I will discuss this pair production process in detail.

The most common example of pair production involves the interaction of a γ-ray (E ≥ 1.02 MeV) with the electric field of an atom, which results in the emission of an electron-positron pair [2].

$$\gamma \rightarrow e^- + e^+ \quad (\text{VI. } 1)$$

I will introduce a physical process that I designate the "Ockham" (or "Occam") Process, to describe electron-positron pair production. I propose further that the Ockham process operates at the upper end of the energy scale, near the Planck energy, where I have postulated that the first quarks appeared (near 6.9625×10^{17} GeV), nearly 21 orders of magnitude greater than the γ-ray energy responsible for electron-positron production. Therefore, I emphasize an essential similarity between the processes

$$\gamma \rightarrow e^- + e^+ \quad (\text{VI. } 1)$$

$$g\bar{g} \rightarrow q + \bar{q} \quad (\text{VI. } 2)$$

despite the factor of 10^{21} (in energy) that separates the two processes. The γ-ray and the gluons ($g\bar{g}$) are each assumed to be massless bosons that "condense" into massive particle pairs, either lepton pairs (Eq. VI.1) or quark pairs (Eq. VI. 2).

Recall the four fundamental forces: gravity, electromagnetic, strong, and weak. Three forces, gravity, electromagnetic, and weak act upon the leptons (electrons, muons, taus), whereas all four forces act upon the quarks.

I will use the example of electron-positron pair production to illustrate the Ockham process. I will then apply the Ockham process to heavy quarks near the Planck mass.

Consider equation VI.1, and write an equation that describes the transfer of energy that occurs when the photon (γ-ray) is converted into an electron-positron pair.

$$E_T = E_m + E_G + E_Q \quad (\text{VI. } 3),$$

where E_T is the total energy of the photon (hν), and E_m , E_G , and E_Q are the energies contained in the electron-positron pair. E_m is composed of the rest mass energy of the pair ($2m_0 c^2$) plus the kinetic energy of the pair. E_G is the gravitational energy of the pair. Q is the symbol for electric charge (in Coulombs) and E_Q is the energy contained in the electric field of the pair.

It occurred to me there might exist a rough analogy between pair production and the condensation of steam into water droplets. In other words, pair production is like a phase change, where bosons become fermions, just as a gas (steam) turns into liquid (water droplets). Likewise, I felt the analogy would also apply to the conversion of gluons into quark pairs, near the Planck energy.

It also seemed reasonable that the leptons and quarks could assume a small, but finite spherical shape with mass m, surface charge density σ , and spin ± 1/2 (since leptons and quarks are fermions). I have designated these spherical structures "Occams" (choosing the simpler spelling). Therefore, I assume that lepton pairs and quark pairs are Occam pairs, produced by the Ockham (or Occam) process. The Occam representation for the electron-positron dipole is shown in Fig. VI. 1. The Occam diameter is d and the dipole spacing is r.

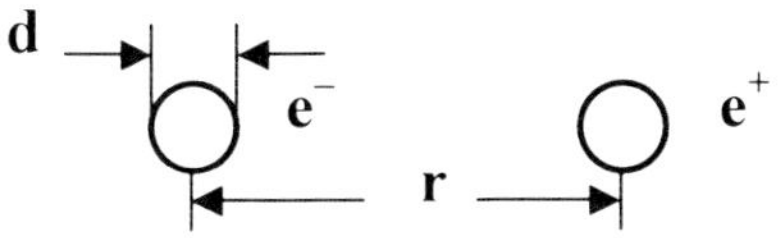

Fig. VI.1
Occam representation of the electron-positron dipole.

I return to the energy equation (VI. 3) to discuss each of the energy terms in detail. Equation (VI.3) applies only to leptons because quarks require the addition of another term E_C , the "color" energy. The color energy will be discussed later when we consider the Occam process for quarks. The pair production equation (VI.1) describes the conversion of a boson (γ) into a lepton pair ($e^- + e^+$). Note that ($e^- + e^+$) is a dipole, with both gravitational and electromagnetic dipole fields. As the Occam process progresses, $2m_0$ and Q increase from **zero** to their final values of the electron and positron rest masses and charges (e^-, e^+), respectively.

Equation (VI.3) contains four energy terms. On the left side is E_T , the original photon energy. The right side contains three terms, E_m , E_G , and E_Q , where E_m is the energy contained in the electron and positron masses, E_G is the gravitational dipole energy and E_Q is the electric dipole energy. The formulas for the energy terms are:

$E_T = h\nu$ (VI.4), where
h = Planck's constant
ν = photon frequency

$E_m = 2m\,c^2 = 2m_0\,c^2 + E_K$ (VI.5), where
2m = total mass of the electron-positron pair
$2m_0$ = rest mass of the electron-positron pair
E_K = kinetic energy of the electron-positron pair

$E_G = G_N m^2 / r$ (VI.6), where
G_N = Newtonian gravitational constant [3]
m = electron mass
r = dipole spacing
d = Occam diameter

$E_Q = k Q^2 / r$ (VI.7), where
$k = 1 / 4\pi \varepsilon_0$
ε_0 = permittivity of free space [4]
Q = e, the electron charge, then
$E_Q = Q^2 / 4\pi \varepsilon_0 r = e^2 / 4\pi \varepsilon_0 r$ (VI.8)

The electron and positron shown in Fig. VI. 1 exhibit both gravitational and electromagnetic attractions across the distance r. The energies E_G and E_Q are the gravitational and electromagnetic energy expenditures required to pull e^- and e^+ apart, increasing the separation r to infinity. However, if r were allowed to approach **zero**, E_G and E_Q would approach infinity. Therefore, there is a good physical reason for assuming that e^- and e^+ have a finite size (d), namely, to prevent r from becoming zero and E_G and E_Q approaching infinity.

We can examine the energetics of the Occam process when we substitute the expressions for E_T , E_m , E_G , and E_Q (Eqs. VI. 4, 5, 6, 8), into Eq (VI. 3):

$$h\nu = 2 m c^2 + G_N m^2 / r + Q^2 / 4\pi \varepsilon_0 r \quad \text{(VI.9)}$$

As the photon is converted into the Occam pair, m, Q, r, and d, all increase from their initial values of zero. Note that the quantities (m^2 / r) and (Q^2 / r) must remain finite as m, Q, and r approach zero, otherwise we would have nonphysical, infinite energies.

For the electron-positron pair, the Occam process proceeds as follows:

(1) The photon interacts with the electric field surrounding the nucleus of the target atom.

(2) The photon momentum is transferred to the target atom through electromagnetic interactions with the atomic electrons.

(3) The Occam process is initiated with the nucleation of the e^- and e^+ Occams.

(4) As the photon energy is transferred to the e^- and e^+ Occams, m, Q, r, and d increase until the total photon energy is transferred and partitioned among the three energy terms on the right-hand side of Eq. (VI.9). 2 m increases until the endpoints are realized, i.e. 2 m = 2 m_0 + E_K (m_0 = electron rest mass, E_K = kinetic energy of pair) and Q = e, e = electron charge).

(5) If E_K is sufficiently large, the gravitational and electric dipole attractions are overcome, the dipole energies decrease to zero, and e^- and e^+ fly apart and continue unimpeded until they interact with other bits of matter. The e^- is absorbed and since e^+ is antimatter, it undergoes an annihilation reaction with another nearby e^-.

For a quark pair, the Occam process proceeds just as described for the electron-positron pair (steps 1 – 5 above) with the exception that the energy equation (VI.3) is modified to include a "color" energy term, giving

$$E_T = E_m + E_C + E_G + E_Q \qquad \text{(VI. 10).}$$

As will be seen later, the color energy (E_C) dependency goes as "r", rather than "1/r", the dependency exhibited in the E_G and E_Q terms [5].

References

[1] K. A. Olive and J. A. Peacock, *Big-Bang Cosmology, Particle Physics Booklet,* U. S. Department of Energy, Pub, 72, July, 2004, p. 223.

[2] R. D. Evans, *The Atomic Nucleus*, McGraw-Hill Book Company, 1955, p. 702.

[3] Particle Data Group, *Particle Physics Booklet,* U. S. Department of Energy, Pub, 72, July, 2004, p. 6.

[4] Particle Data Group, *Particle Physics Booklet,* U. S. Department of Energy, Pub, 72, July, 2004, p. 4.

[5] C. DeTar, *Quantum Flavordynamics, Quantum Chromodynamics, and Unified Field Theories*, edited by K. T. Mahanthappa and J. R. Randa, Plenum, New York, 1980, p. 411.

VII The Ockham Process – Part 2

Synopsis

The energy expressions for the electron-positron pair (Eq. VI. 3) and the quark pair (Eq. VI. 10) will be used to derive the Occam properties of each pair.

Analysis

Following the derivation of the energy terms in equations (VI. 3) and (VI. 10), we will nondimensionalize the equations [1] and derive the stability conditions for the lepton and quark pairs. Equations (VI. 3) and (VI. 10), for the lepton pair and the quark pair, respectively, are the same with the exception of the color term in (VI. 10). Therefore, we will consider the more general quark pair equations first, then specialize later to the lepton pair. Repeating Eq. (VI. 10), we have:

$E_T = E_m + E_C + E_G + E_Q$ (VII. 1),

where E_T, E_m, E_C, E_G, and E_Q are the total, mass, color, gravitational, and electromagnetic energies, respectively. Repeating the relevant parts of Eqs (VI. 4, 5, 6, 8) and adding the color term, we have:

$E_T = h\nu$ (VII.2)
$E_m = 2\,m\,c^2 = 2\,m_0\,c^2 + E_K$ (VII.3)
$E_C = (\text{constant}) \bullet (r)$ (VII.4)
$E_G = G_N\,m^2 / r$ (VII.5)
$E_Q = Q^2 / 4\pi\,\varepsilon_0\,r$ (VII.6)
where
h = Planck's constant
ν = gluon frequency
m = quark total mass

m_0 = quark rest mass
c = speed of light
E_K = kinetic energy of the quark pair
r = dipole length
d = Occam diameter
G_N = Newtonian gravitational constant
Q = quark charge
ε_0 = permittivity of free space

Consider the Occam pair for quarks, a dipole that consists of a quark and antiquark (Fig. VII. 1). This figure is essentially the same as the lepton pair, Fig. VI. 1. In Chapter VI, I described the cooling of the gluon ensemble, followed by the "condensation" of the topmost quarks via the Occam process. As the gluons cooled, the gluon energies decreased and the gluon wavelengths increased. Equation VII. 1 will be used to calculate the gluon wavelength at which quark condensation occurs.

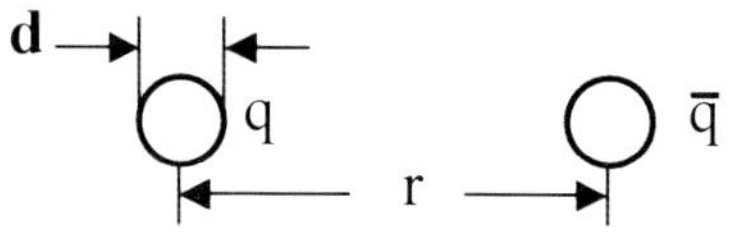

Fig. VII.1
Occam representation of the quark-antiquark dipole.

This critical wavelength is calculated by expressing the four energy terms in Eq. VII. 1, that constitute E_T, as functions of gluon wavelength λ and dipole length **r**. We find that $E_T(\lambda, \mathbf{r})$ has a minimum, which is easily determined by elementary differential calculus, i.e. by the equation

$$d E_T(\lambda, \mathbf{r}) / d\lambda = 0 \qquad \text{(VII. 7)},$$

where λ is the minimization variable and **r** is treated as a free parameter, which we will determine later. This minimum defines the gluon wavelength at which the transformation from gluon to quark pair occurs.

Next, we express the four energy terms in Eq. (VII. 7) as functions of λ and evaluate Eq. (VII. 7). As we proceed with these evaluations, we will nondimensionalize Eq. (VII. 7). We then obtain an expression for the value of λ that satisfies (VII. 7) in terms of the threshold wavelength λ_0, where λ_0 is the wavelength for which $E_G = E_Q$. λ_0 is actually the first approximation discussed in Chapter II, where we recall from Chapter II, that we assumed that the quark pair condensed from a gluon with energy $2\ m\ c^2 = h\ \nu = h\ c\ /\ \lambda_0$. Then, $\lambda_0 = h\ /\ 2\ m\ c$, where $m = (k / G_N)^{1/2}\ Q = 6.9625 \times 10^{17}$ GeV (II.2).

Referring to Fig. VII. 1, we proceed as follows, introducing the following nondimensional quantities:

$\rho = r / \lambda \ ; \quad \delta = d / \lambda \ ; \quad x = \lambda / \lambda_0$ (VII. 8a)

$r = \rho \lambda \ ; \quad d = \delta \lambda \ ; \quad \lambda = \lambda_0 x$ (VII. 8b)

Evaluate the four energy terms in Eq. (VII. 1):

$$E_m = 2\ m\ c^2 = h\nu = h\ c / \lambda = h\ c /\ \lambda_0\ x = (h\ c /\ \lambda_0)\ x^{-1} = M\ x^{-1} \quad \text{(VII.9)}$$

$$E_C = (\text{constant}) \bullet r = (\text{constant}) \bullet \rho\ \lambda = (\text{constant}\ \rho\ \lambda_0) \bullet x = C\ x \quad \text{(VII.10)}$$

$$E_G = G_N m^2 / r \quad ; \quad m = h / 2\ \lambda\ c$$

$$E_G = (G_N / r) \bullet (h^2 / 4\ \lambda^2 c^2) = (G_N / \rho\ \lambda) \bullet (h^2 / 4\ \lambda^2 c^2) = G_N h^2 / 4\ \rho\ c^2 \lambda^3 = G_N h^2 / 4\ \rho\ c^2\ \lambda_0^3 x^3 = (G_N h^2 / 4\ \rho\ c^2\ \lambda_0^3)\ x^{-3} = A\ x^{-3} \quad \text{(VII.11)}$$

Recall Fig. VII.1. The total Occam charge Q is given by the surface charge density σ , times the Occam surface area:

$$E_Q = Q^2 / 4\pi\, \varepsilon_0\, r = (\pi\, \sigma\, d^2)^2 / 4\pi\, \varepsilon_0\, r = \pi\, \sigma^2\, d^4 / 4\, \varepsilon_0\, r =$$
$$\pi\, \sigma^2 (\delta\, \lambda)^4\ / \ 4\, \varepsilon_0\, \rho\, \lambda = \pi\, \sigma^2\, \delta^4\, \lambda^3 / 4\, \varepsilon_0\, \rho\ = \pi\, \sigma^2\, \delta^4\, (\lambda_0\, x)^3 / 4\, \varepsilon_0\, \rho =$$
$$(\pi\, \sigma^2\, \delta^4\ \lambda_0^3 / 4\, \varepsilon_0\, \rho)\, x^3\ = B\, x^3 \qquad \text{(VII.12)}$$

Substituting (VII.9 – 12) into (VII.1), we have:

$$E_T = M\, x^{-1} + C\, x + A\, x^{-3} + B\, x^3 \qquad \text{(VII.13),}$$

where x is the nondimensional wavelength and

$M = h\, c / \lambda_0$; $C = \text{constant} \bullet \rho\ \lambda_0$

$A = G_N\ h^2 / 4\ \rho\ c^2\ \lambda_0^3$; $B = \pi\, \sigma^2\, \delta^4\ \lambda_0^3\ / \ 4\, \varepsilon_0\ \rho$

M, C, A, and B all having the units of energy (Joules).

Equation (VII.13) is now expressed in terms of the nondimensional wavelength x and the nondimensional dipole length ρ . Therefore, Eq. (VII. 7) becomes

$$d\, E_T\, (x, \rho) / d\, x = 0 \qquad \text{(VII. 14),}$$

and ρ is a temporarily free parameter, which appears in the terms C, A, and B, in Eq. (VII. 13). Note that of the four terms (VII.9) - (VII.12), only the first is not a dipole energy term. We can fix ρ by assuming that the initial gluon energy is divided equally between the mass energy term and the sum of the dipole energy terms, $C\, x + A\, x^{-3} + B\, x^3$.

$$\therefore\ M\, x^{-1} = C\, x + A\, x^{-3} + B\, x^3 \qquad \text{(VII.15).}$$

Evaluate Eq. (VII. 14):

$d\,E_T\,(x, \rho)\,/\,d\,x = 0$ (VII. 14), then

$$d\,E_T\,(x, \rho)\,/\,d\,x = d\,(M\,x^{-1} + C\,x + A\,x^{-3} + B\,x^3)\,/\,d\,x$$

$$= -\,M\,x^{-2} + C - 3\,A\,x^{-4} + 3\,B\,x^2 = 0 \qquad \text{(VII. 16).}$$

Multiply Eq. (VII. 15) by x^3:

$$M\,x^2 = C\,x^4 + A + B\,x^6 \qquad \text{(VII. 17).}$$

Multiply Eq. (VII. 16) by x^4:

$$M\,x^2 = C\,x^4 - 3\,A + 3\,B\,x^6 \qquad \text{(VII. 18).}$$

Subtract Eq. (VII. 17) from Eq. (VII. 18):

$$2\,B\,x^6 - 4\,A = 0$$

$$x^6 = 2\,A\,/\,B$$

$$x = (2\,A\,/\,B)^{1/6} \qquad \text{(VII. 19)}$$

In Chapter II, we made the assumption that $E_G = E_Q$, in order to determine the gluon wavelength λ_0, for which the topmost quark pair condensed. In terms of the nondimensional wavelength $x = \lambda\,/\,\lambda_0$, the equality $E_G = E_Q$ becomes

$$E_G = A\,x^{-3} = B\,x^3 = E_Q\,.$$

For the first wavelength approximation in Chapter II, $\lambda = \lambda_0\,x = 1$, and $A = B$. For the second wavelength approximation, we again let $A = B$. When $A = B$ in Eq. (VII. 19), we obtain the second approximation for x, namely $x = 2^{1/6} = 1.122462 > 1$. Since $E_G = A\,x^{-3}$ and $E_Q = B\,x^3$, then $E_Q\,/\,E_G = B\,x^6\,/\,A$. When $A = B$, $E_Q\,/\,E_G = x^6 = 2$, rather than $E_Q\,/\,E_G = 1$. Therefore, the total energy E_T is minimized when $x = 2^{1/6}$, rather than $x = 1$. Finally, we have $\lambda_{min} = 2^{1/6}\,\lambda_0 = 1.122462\;\lambda_0$, which is roughly 12 % greater than the first approximation from Chapter II.

Recall the charge states for leptons and quarks:

For leptons, $Q = -e$
For an up quark, $Q = +2e/3$
For a down quark, $Q = -e/3$

If f is the fraction of the electron charge, then $Q = f\,e$, where $f = \pm 1/3, \pm 2/3, \pm 1$,
then $E_Q = Q^2/4\pi\varepsilon_0 r = f^2 e^2/4\pi\varepsilon_0 r$.

The next step is the determination of the masses of the topmost quarks. Recall the expressions for E_G and E_Q:
$E_G = G_N m^2/r$; $E_Q = Q^2/4\pi\varepsilon_0 r = f^2 e^2/4\pi\varepsilon_0 r$

The fine structure constant α is given by [2]:
$\alpha = e^2/4\pi\varepsilon_0 h_r c$, then $\alpha h_r c = e^2/4\pi\varepsilon_0$. Since $E_Q = 2E_G$, $2G_N m^2/r = f^2 e^2/4\pi\varepsilon_0 r$.
Dropping the r's and substituting the expression for $e^2/4\pi\varepsilon_0$, we have $2G_N m^2 = f^2 \alpha h_r c$; $m^2 = f^2 \alpha h_r c / 2G_N$, and $m = f\sqrt{(\alpha h_r c / 2G_N)}$
The Planck mass is given by [3] $m_P = \sqrt{(h_r c / G_N)}$, from which we finally obtain
$m = f\sqrt{(\alpha/2)}\, m_P$ (VII. 20).

Substituting the following into Eq. (VII. 20):
$f = 2/3$ for U_R, and $m(D_L) = m(U_R)/4$
$\alpha = 1/137.036$

$m_P = 1.2209 \times 10^{19}$ GeV / c^2, we find
$m(U_R) = 4.9165 \times 10^{17}$ GeV / c^2
$m(D_L) = 1.2291 \times 10^{17}$ GeV / c^2

We now solve for the dipole spacing r :

$E_T = 2\ m\ c^2$

$M = C + A + B$

$M = m\ c^2$

$C = A = B \Rightarrow$

$3\ G_N m^2 / r = m\ c^2 = h\ c / \lambda$

$3\ G_N m^2 = (r / \lambda)\ h\ c$

$r / \lambda = (3\ G_N / hc)\ m^2 = (3\ G_N / hc) \bullet (f^2 \alpha\ m_P^2 / 2) = (3\ G_N / hc) \bullet (f^2 \alpha / 2) \bullet (h_r\ c / G_N)$

$r / \lambda = 3\ f^2 \alpha / 4\pi$ (VII. 21)

References

[1] H. L. Langhaar, *Dimensional Analysis and Theory of Models,* John Wiley and Sons, New York, 1951, Fifth Printing, 1962.

[2] Particle Data Group, *Particle Physics Booklet,* U. S. Department of Energy, Pub, 72, July, 2004, p. 4.

[3] Particle Data Group, *Particle Physics Booklet,* U. S. Department of Energy, Pub, 72, July, 2004, p. 6.

VIII The Ockham Process – Part 3

Synopsis

We modify equations (VII. 13) and (VII. 14) in order to examine the Occam process for the electron-positron pair.

Analysis

Equations (VII. 13) and (VII. 14) simplify considerably for the electron-positron pair. Since leptons do not interact with color forces, C = 0, and for the electron-positron pair, A can be shown to be negligible with respect to B, as follows:

$E_G = G_N m^2 / r = A x^{-3}$ (VII.11)

$E_Q = Q^2 / 4\pi \varepsilon_0 \ r = B x^3$ (VII.12)

Substituting,
$G_N = h_r c / m_P^2$; $E_G = G_N m^2 / r = G_N m_e^2 / r = (h_r c / m_P^2) \bullet (m_e^2 / r)$
$E_G = (h_r c / r) \bullet (m_e / m_P)^2$

$E_Q = Q^2 / 4\pi \varepsilon_0 r = e^2 / 4\pi \varepsilon_0 r$
$e^2 / 4\pi \varepsilon_0 = \alpha h_r c$; $E_Q = \alpha h_r c / r$

$E_G / E_Q = (h_r c / r) \bullet (m_e / m_P)^2 \bullet (r / \alpha h_r c) = (m_e / m_P)^2 / \alpha$
$\alpha = 1/ 137.036$
$m_e = 9.10938 \times 10^{-31}$ kg
$m_P = 2.17645 \times 10^{-8}$ kg, then
$E_G / E_Q = (m_e / m_P)^2 / \alpha = 2.40057 \times 10^{-43}$
Clearly, E_G is negligible with respect to E_Q .

Since $E_G = A\,x^{-3}$, $E_Q = B\,x^3$, and $x \cong 1$,
$E_G / E_Q = (A\,x^{-3})/(B\,x^3) = (A/B)\,x^{-6} \cong A/B$
$\therefore$ A is clearly negligible with respect to B.

Now, since C = A = 0, Eq. (VII.13) becomes
$E_T = M\,x^{-1} + B\,x^3$ (VIII.1)

As before, we differentiate the total energy (VIII.1) with respect to x:
$d\,E_T\,(x, \rho) / d\,x = d\,(M\,x^{-1} + B\,x^3) / d\,x = -\,M\,x^{-2} + 3\,B\,x^2 = 0$
(VIII. 2).

Multiply (VIII. 2) by x^2 and solve for x:
$-\,M + 3\,B\,x^4 = 0$
$x^4 = M/3B$; $x = (M/3B)^{1/4}$ (VIII. 3).

Recall that C = A = 0. Then Eq (VII. 15) becomes:
$M\,x^{-1} = B\,x^3$
Since $x \cong 1$, $M \cong B$, then $x = (1/3)^{1/4} = 0.7598$

We now solve for the dipole spacing r :
$M = m_e\,c^2$; $B = f^2\,e^2 / 4\pi\,\varepsilon_0\,r = e^2 / 4\pi\,\varepsilon_0\,r$

$e^2 / 4\pi\,\varepsilon_0 = \alpha\,h_r\,c$; then
$m_e\,c^2 = \alpha\,h_r\,c/r$, then
$r_e = \alpha\,h_r / (m_e\,c) = \alpha\,h / (2\pi\,m_e\,c)$
From [1], the Compton wavelength is
$\lambda_e = h / (m_e\,c) = 2\pi \bullet 3.861593 \times 10^{-13}$ m $= 2.4263 \times 10^{-12}$ m
$r_e / \lambda_e = (\alpha\,h / 2\pi\,m_e\,c) \bullet (m_e\,c) / h = \alpha / 2\pi$ (VIII. 4).

Comparing equations (VII. 21) and (VIII. 4), we note that the topmost quarks and the electron have similar ratios for (r / λ):
For an up quark ($f^2 = 4/9$), $r / \lambda = \alpha / 3\pi$
For a down quark ($f^2 = 1/9$), $r / \lambda = \alpha / 12\pi$
For the electron ($f^2 = 1$), $r / \lambda = \alpha / 2\pi$

From the Particle Physics Booklet [1], we note that the classical electron radius is $r_e = 2.81794 \times 10^{-15}$ m. In the present analysis, we identify r_e with the electron-positron dipole length, not one-half of the Occam diameter d. Quigg [2], states that "all known [leptons] are structureless at the current limits of resolution of 1×10^{-16} cm", or 1×10^{-18} m.

From Eq. (VIII. 4), we clearly see that $r_e >> 1 \times 10^{-18}$ m :
$r_e = (\alpha / 2\pi) \bullet \lambda_e = 2.4263 \times 10^{-12}$ m / $(2\pi \bullet 137.036) =$
2.8179×10^{-15} m $>> 1 \times 10^{-18}$ m.
This result also clearly implies that $d_e << r_e$, provided the condition $d_e < 1 \times 10^{-18}$ m $<< r_e$ holds true.

References

[1] Particle Data Group, *Particle Physics Booklet,* U. S. Department of Energy, Pub, 72, July, 2004, p. 4.

[2] C. Quigg, *Gauge Theories of the Strong, Weak, and Electromagnetic Interactions,* Westview Press, 1997, p. 3.

IX The Ockham Process – Part 4

Synopsis

Estimates for the Occam diameters d_m and d_e are obtained by relating d_m to r_m, and relating d_e to λ_e using dimensional analysis.

Analysis

The previous analyses have produced estimates for the masses of the topmost quarks U_R and D_L, and the dipole lengths r_m and r_e, for the topmost quark pairs and the electron-positron pair, respectively.

The topmost quark mass and dipole length are (VII. 20), (VII. 21):

$$m = f\sqrt{(\alpha/2)}\, m_P \qquad \text{(IX. 1)}$$

$$r_m / \lambda_m = 3 f^2 \alpha / 4\pi \qquad \text{(IX. 2)}$$

$$\lambda_m = 2\pi h_r / m c \qquad \text{(IX. 3)}$$

$$r_m = 3 h_r f \sqrt{(\alpha/2)} / m_P c \qquad \text{(IX. 4)}$$

Substituting the quantities:
$h_r = 1.054572 \times 10^{-34}$ J s
$\alpha = 1/137.036$
$m_P = 2.17645 \times 10^{-8}$ kg
$c = 2.99792 \times 10^{8}$ m/s
$f = 2/3$

$m = f\sqrt{(\alpha/2)}\, m_P = 8.7645 \times 10^{-10}$ kg
$\lambda_m = 2\pi h_r / m c = 2.5218 \times 10^{-33}$ m
$r_m = 3 h_r f \sqrt{(\alpha/2)} / m_P c = 1.9526 \times 10^{-36}$ m

Return to Fig. VII. 1 and consider the ratio d / r , which can vary from 1 to 0 as d decreases with respect to r. We will consider two values for the ratio d_m / r_m , i.e. 1 and 1 / 10 , and obtain estimates for d_e , the Occam diameter for the electron.

The mass density (ρ) for an Occam (either U_R, D_L, or e^-) is given by:
$\rho = m / V$, where
m = Occam mass
V = Occam volume
$V = (4\pi / 3) \bullet (d /2)^3 = \pi d^3 / 6$
$\rho = 6 m / \pi d^3$
$m = h c / \lambda_m$, where
h = Planck's constant
c = speed of light
λ_m = deBroglie wavelength of the Occam

For the calculations that follow, m refers to U_R, the topmost quark. We propose the following relationship between Occam diameters and deBloglie wavelengths:
$d / d_m = (\lambda / \lambda_m)^n$ (IX. 5),
where we will use dimensional analysis to solve for n.

Eq. (IX. 5) becomes
$d = (d_m / \lambda_m{}^n) \lambda^n = D \lambda^n$, then
$\rho = m/V = (h c / \lambda) \bullet (6 /\pi d^3) = (6 h c /\pi) \bullet (1 / \lambda d^3)$
$= (6 h c /\pi) \bullet (1 / \lambda) \bullet (1 / D^3 \lambda^{3n}) = (6 h c /\pi D^3) \bullet (1 / \lambda^{3n+1})$
(IX. 6)

If we also assume that
$\rho = m/V \propto 1 / \lambda^3$, we can solve for n:
$\rho = (6 h /\pi c D^3) \bullet (1 / \lambda^{3n+1}) \propto 1 / \lambda^3$ (IX. 7)

Equating the exponents of λ , we have:
$3n + 1 = 3$
$n = 2/3$

Letting $d = d_e$ and $\lambda = \lambda_e$, we have
$d_e / d_m = (\lambda_e / \lambda_m)^{2/3}$.
$\lambda_e = 2.4263 \times 10^{-12}$ m, $\lambda_m = 2.5218 \times 10^{-33}$ m, and we examine two possibilities for d_m / r_m , 1 and 1 / 10 .
For $d_m / r_m = 1$, $d_m = 1.9526 \times 10^{-36}$ m
For $d_m / r_m = 1/10$, $d_m = 1.9526 \times 10^{-37}$ m
$d_e = d_m (\lambda_e / \lambda_m)^{2/3}$.
Substituting the range: 1.9526×10^{-36} m – 1.9526×10^{-37} m, for d_m , we obtain:
$d_e = 1.9030 \times 10^{-22}$ m – 1.9030×10^{-23} m .
We have seen experimentally [1] that $d_e < 10^{-18}$ m, therefore, our estimate:
$d_e = 1.9030 \times 10^{-22}$ m – 1.9030×10^{-23} m, clearly fits the requirement $d_e < 10^{-18}$ m .

Reference

[1] C. Quigg, *Gauge Theories of the Strong, Weak, and Electromagnetic Interactions,* Westview Press, 1997, p. 3.

X The Big-Bang and Hyperneutrons – Part 1

Synopsis

Returning to the initial conditions discussed in Chapter V, we examine the cooling of the gluon ensemble and the production of the topmost quarks via the Occam process.

The cooling gluons are postulated to transform initially via a six-body interaction designated {6g1}, followed by further cooling and another six-body interaction designated {6g2}. The {6g1} interaction results from a straightforward combination of two g_7 gluons. The {6g2} interaction is more complicated, involving precursor interactions of the gluon composites $\sqrt{(3)}\ g_7 \pm g_8$, which result eventually in the {6g2} interaction.

The {6g1} and {6g2} interactions each produce the quarks U_R and D_L, and their antiquarks. However, the energy distributions among the U_R and D_L quarks produced by the {6g1} and {6g2} interactions are significantly different. Finally, the U_R and D_L quarks coalesce into two distinct sets of "hyperneutrons" and "antihyperneutrons", with two distinct and differing decay behaviors, leading to Regular Matter (RM) in one case, and Dark Matter (DM) in the other case.

Analysis

The following describes the process details for the transformation of g_7 and g_8 gluons into U_R and D_L quarks. As I discussed in Chapters VI - IX, quark transformations are analogous to ordinary pair production, in which a boson (photon) is converted into a fermion (electron) and an antifermion (positron). In an analogous process, the bosons g_7 and g_8 are converted into fermions (U_R quark and D_L quark) and antifermions ($\bar{U}_R$ antiquark and $\bar{D}_L$ antiquark).

As stated above, the {6g1} process results from the straightforward interaction of two g_7 gluons. The second process also involves the collective actions of six gluons and is designated {6g2}. However, the {6g2} process is preceded by a sequence of interactions involving the composite gluons $\sqrt{(3)}\, g_7 + g_8$ and $\sqrt{(3)}\, g_7 - g_8$.

The amounts of Regular Matter (RM) and Dark Matter (DM) produced, following the Big-Bang, are postulated to be directly proportional to the interaction cross sections for the g_7 gluons and the ($\sqrt{(3)}\, g_7 \pm g_8$) composite gluons. As noted in Chapter V,

$$g_7 = (2R\bar{R} - B\bar{B} - G\bar{G}) / \sqrt{(6)}$$

$$g_8 = (B\bar{B} - G\bar{G}) / \sqrt{(2)}$$

We postulate that U_R ($Q = +2/3$) and $\bar{U}_R$ ($Q = -2/3$) originate from the $2R\bar{R}$ fraction of g_7 and a pair of D_L ($Q = -1/3$) and a pair of $\bar{D}_L$ ($Q = +1/3$) originate from the $B\bar{B}$ and $G\bar{G}$ fractions of g_7, resulting in the triples { U_R , D_L , D_L }and{ $\bar{U}_R$, $\bar{D}_L$, $\bar{D}_L$ }. For the gluon composite $\sqrt{(3)}\, g_7 \pm g_8$, we have:

$$\sqrt{(3)}\, g_7 + g_8 = (2R\bar{R} - B\bar{B} - G\bar{G}) / \sqrt{(2)} + (B\bar{B} - G\bar{G}) / \sqrt{(2)}$$

$$= (2R\bar{R} - 2G\bar{G}) / \sqrt{(2)}$$

$$\sqrt{(3)}\, g_7 - g_8 = (2R\bar{R} - B\bar{B} - G\bar{G}) / \sqrt{(2)} - (B\bar{B} - G\bar{G}) / \sqrt{(2)}$$

$$= (2R\bar{R} - 2B\bar{B}) / \sqrt{(2)}$$

We postulate that the $\sqrt{(3)}\, g_7 \pm g_8$) composites behave in a fashion similar to the g_7 gluon, in that triples { U_R , D_L , D_L }and{ $\bar{U}_R$, $\bar{D}_L$, $\bar{D}_L$ } arise from each of the composites $\sqrt{(3)}\, g_7 + g_8$ and $\sqrt{(3)}\, g_7 - g_8$, with U_R and $\bar{U}_R$ originating from $2R\bar{R}$ and a pair of D_L and $\bar{D}_L$ originating from either $2B\bar{B}$ or $2G\bar{G}$. We now examine and compare the {6g1} and {6g2} interactions.

We will use simple approximations, involving only the deBroglie wavelengths for the gluons, to estimate the interaction cross sections for {6g1} and {6g2}. The $2\bar{R}R$ fraction of g_7 is postulated to contain four times the energy of the $B\bar{B}$ fraction or the $G\bar{G}$ fraction. Since $E = hc / \lambda$, the deBroglie wavelength of $2R\bar{R}$ is 1/4 the deBroglie wavelength of $B\bar{B}$ or $G\bar{G}$.

We employ the normalized deBroglie wavelengths:

$$2R\bar{R} / \sqrt{(6)} = 1$$
$$B\bar{B} / \sqrt{(6)} = G\bar{G} / \sqrt{(6)} = 4 ,$$

then the cross section for g_7 can be depicted as shown in Fig. X.1, where the gluon cross sections are represented in red, blue, and green, and the g_7 "footprint" is the shaded area.

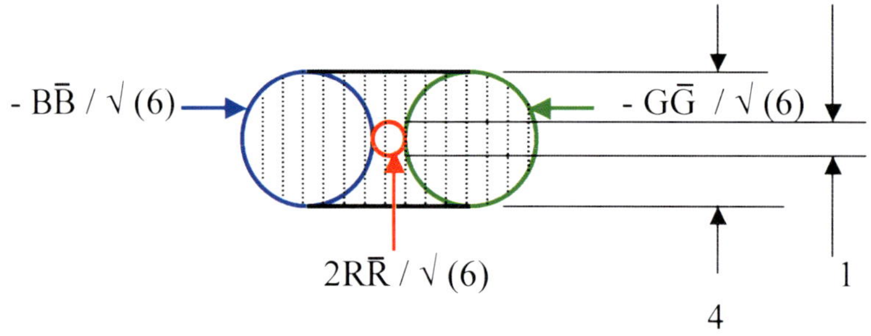

Fig. X.1
g_7 gluon cross section

Next, we will examine two different quark transformation processes, each involving the collective actions of six gluons and designated the {6g1} and {6g2} processes, respectively. Both processes are initiated with equal numbers of right and left-handed gluons. Each process is momentum, spin, and CP conserving. Each of the two processes produces the following ensembles of right and left-handed quarks and antiquarks; 2 each of {U_R , D_L , D_L , $\overline{U}_R$, $\overline{D}_L$, $\overline{D}_L$}. However, the distributions of right and left-handed kinetic energies are quite different for each ensemble. We postulate that these energy differences are responsible for the production of RM by the {6g1} process and the production of DM by the {6g2} process.

We begin with the {6g1} process. Quark formation is initiated as the g_7 gluons cool to the quark formation threshold temperature. The {6g1} process is initiated by gluon interactions with virtual dipoles $Q^+ \overline{Q^+}$ (Fig. X.2). Three right-handed gluons interact with the plus half of the virtual dipole and three left-handed gluons interact with the minus half of the virtual dipole.

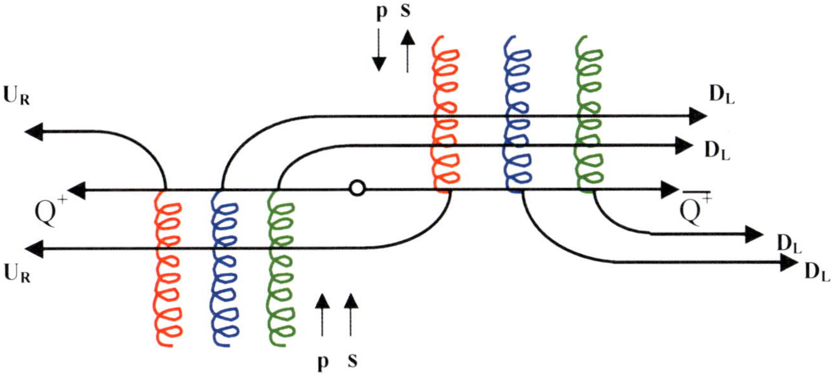

Fig. X.2
{6g1} process. g_7 gluons interact collectively with virtual dipole and Lorentz forces to produce U_R and 2 D_L quarks. For clarity, antiquarks are not shown.

The $2R\bar{R}$ gluon produces a U, $\bar{U}$ pair and the $B\bar{B}$ and $G\bar{G}$ gluons each produce a D, $\bar{D}$ pair. The Lorentz forces, that result from the motions of the dipole charges, drive the positively charged U quarks to the left and the negatively charged D quarks to the right. In similar fashion, the negatively charged U antiquarks and the positively charged D antiquarks are driven to the right and left, respectively. For clarity, the antiparticles are not shown in Fig. X.2 .

For the U quarks, momentum and spin are parallel, hence, the Us are right-handed. For the D quarks, momentum and spin are antiparallel, hence, the Ds are left-handed.

The {6g1} and {6g2} processes each produce heavy quarks and antiquarks that combine to form pairs of hyperneutrons $\{U_R , D_L , D_L\}$ and pairs of anti-hyperneutrons $\{\bar{U}_R , \bar{D}_L , \bar{D}_L\}$. Although the {6g1} and {6g2} processes produce apparently identical hyperneutrons and anti-hyperneutrons, the kinetic energy distributions within the {6g1} and {6g2} hyperneutrons and anti-hyperneutrons are quite different.

The {6g2} process is more complex than the {6g1} process. The g_7 gluons that did not participate in the {6g1} process continued to cool and their deBroglie wavelengths increased to $\surd(3)\, g_7$. The {6g2} process began when the $\surd(3)\, g_7$ gluons interacted with g_8 gluons to form the composites $\surd(3)\, g_7 \pm g_8$. Then we have:

$$\surd(3)\, g_7 + g_8 = 2R\bar{R} / \surd(2) - B\bar{B} / \surd(2) - G\bar{G} / \surd(2) + B\bar{B} / \surd(2) - G\bar{G} / \surd(2)$$

$$\surd(3)\, g_7 - g_8 = 2R\bar{R} / \surd(2) - B\bar{B} / \surd(2) - G\bar{G} / \surd(2) - B\bar{B} / \surd(2) + G\bar{G} / \surd(2)$$

Using the normalized deBroglie wavelengths that we used for {6g1}, multiplied by √ (3), we have:

$$2R\bar{R} / \sqrt{(2)} = \sqrt{(3)}$$
$$B\bar{B} / \sqrt{(2)} = G\bar{G} / \sqrt{(2)} = 4\sqrt{(3)},$$

then the cross sections for $\sqrt{(3)}\, g_7 \pm g_8$ can be depicted as shown in Fig. X.3, where, as in Fig. X.1, the gluons are represented in red, blue, and green, and the "footprints" are the shaded areas.

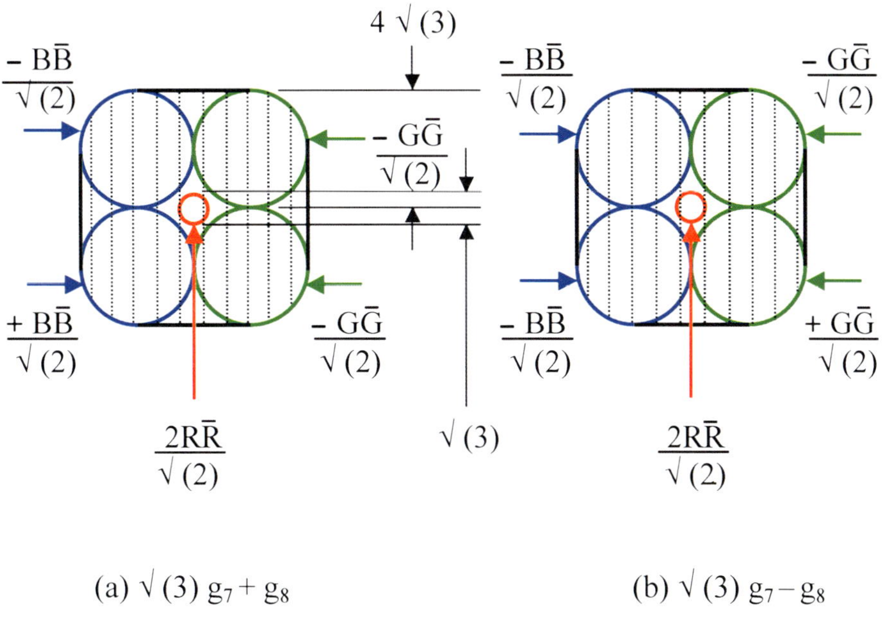

Fig. X.3
$\sqrt{(3)}\, g_7 \pm g_8$ Composite Gluon Cross Sections

However, the {6g2} process differs from {6g1}, in that {6g1} starts and ends with a pair of 3-gluon ensembles, whereas {6g2} starts with two pairs of 5-gluon ensembles and ends with two pairs of 3-gluon ensembles.

The collapse of the 5-gluon ensembles into 3-gluon ensembles proceeds as follows:

$$\sqrt{(3)}\, g_7 + g_8 = 2R\bar{R} / \sqrt{(2)} - B\bar{B} / \sqrt{(2)} - G\bar{G} / \sqrt{(2)} + B\bar{B} / \sqrt{(2)} - G\bar{G} / \sqrt{(2)}$$

$$= 2R\bar{R} / \sqrt{(2)} - G\bar{G} / \sqrt{(2)} - G\bar{G} / \sqrt{(2)}$$

$$\sqrt{(3)}\, g_7 - g_8 = 2R\bar{R} / \sqrt{(2)} - B\bar{B} / \sqrt{(2)} - G\bar{G} / \sqrt{(2)} - B\bar{B} / \sqrt{(2)} + G\bar{G} / \sqrt{(2)}$$

$$= 2R\bar{R} / \sqrt{(2)} - B\bar{B} / \sqrt{(2)} - B\bar{B} / \sqrt{(2)}$$

Fig. X.3 is then transformed into:

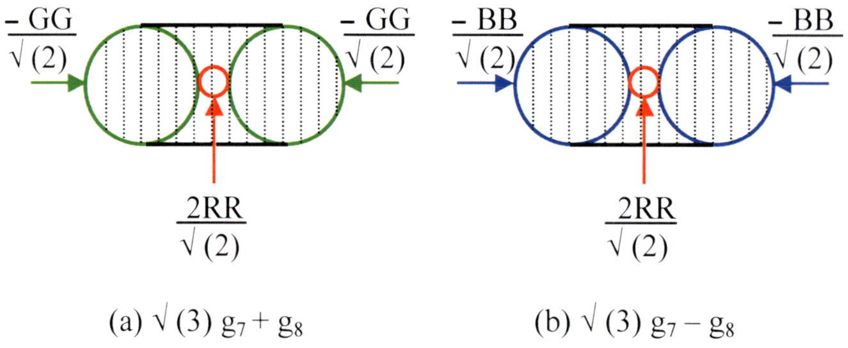

Fig. X.4
$\sqrt{(3)}\, g_7 \pm g_8$ Transformed Gluon Cross Sections

The {6g2} process appears much like the {6g1} process depicted in Fig. X.2, except that unlike Fig. X.2, where each 3-gluon ensemble contains three colors, each 3-gluon ensemble in {6g2} contains two gluons with the same color (Fig. X.5). In other words, if the right-handed 3-gluon ensemble contains blue gluons, then the left-handed 3-gluon ensemble contains green gluons (Fig. X.5(a)). Likewise, if the right-handed gluons are green, then the left-handed gluons are blue (Fig. X.5(b)).

At this juncture, we will consider the different kinetic energy distributions for the hyperneutrons produced by the {6g1} and {6g2} processes. The kinetic energy of a massive particle is given by: $E_K = (m - m_0) c^2$, which is easily shown to be expressible as $E_K = (m v^2/2)[1 + \beta^2/4 + \beta^4/8 + \text{H.O.T}]$, where m = total mass, m_0 = rest mass, v = velocity, c = velocity of light, $\beta = v / c$, and H. O. T. = Higher Order Terms. Therefore, $E_K > (m v^2/2)$ and $(m v^2/2)$ is clearly a lower bound for E_K .

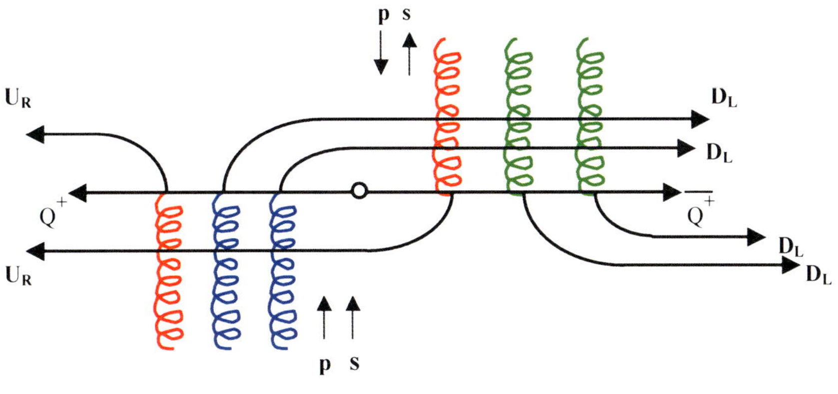

Fig. X.5(a)
{6g2} process. √ (3) $g_7 \pm g_8$ gluon composites interact collectively with virtual dipole and Lorentz forces to produce U_R and 2 D_L quarks. Right-handed gluons are blue and left-handed gluons are green. For clarity, antiquarks are not shown.

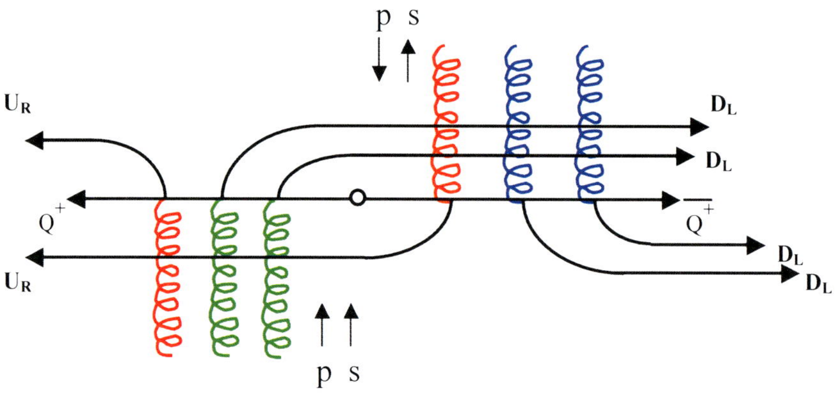

Fig. X.5(b)
{6g2} process. √ (3) $g_7 \pm g_8$ gluon composites interact collectively with virtual dipole and Lorentz forces to produce U_R and 2 D_L quarks. Right-handed gluons are green and left-handed gluons are blue. For clarity, antiquarks are not shown.

For both {6g1} and {6g2} processes, momentum, p = m v, is conserved between quarks and antiquarks, i.e. $p + \overline{p} = 0$. However, for {6g1}, a stronger condition obtains, namely, that momentum is conserved between right and left-handed quarks and antiquarks, i.e.
$p_R + p_L = 0$ and $\overline{p}_R + \overline{p}_L = 0$.

In terms of masses and velocities for the U_R (right-handed) and the two D_L (left-handed) quarks, we have: $m_U v_U + 2 m_D v_D = 0$. Since $m_U = 4 m_D$, we have $4 m_D v_U + 2 m_D v_D = 0$,
and $v_D = -2 v_U$.

The lower-bound kinetic energies for the U_R and D_L quarks are given by the expressions:
$E_{KU} = m_U v_U^2/2$
$E_{KD} = (2)(m_D v_D^2/2) = m_D v_D^2$

Substituting, we have:
$E_{KU} = 4\ m_D\ (v_U^2/2) = 2\ m_D\ (-\ v_D\ /2)^2 = m_D\ v_D^2\ /2$
Therefore: $E_{KU} = E_{KD}\ /\ 2$.

For {6g2} we no longer have $p_R + p_L = 0$ and $\bar{p}_R + \bar{p}_L = 0$, rather we have $p_R + p_L > 0$ and $\bar{p}_R + \bar{p}_L > 0$. Referring to Fig. X.3(a) and Fig. X.3(b), which represent $\surd\ (3)\ g_7 \pm g_8$, we examine the interactions of $\surd\ (3)\ g_7$ and g_8 .

Gluons g_7 and g_8 are postulated to begin with an energy $E_0 >> E_P$, where E_P is the Planck energy. As the gluons cool, they assume energy distributions resembling the black-body energy distribution of an ensemble of photons. It is also reasonable to assume that the gluon momentum vectors are distributed isotropically in space.

Fig. X.3(a) and X.3(b) represent the interactions of gluons $+\ g_8$ and $-\ g_8$ with $\surd\ (3)\ g_7$. A "full physics" treatment of these interactions would involve both the energy and momentum distributions for $\surd\ (3)\ g_7$ and $\pm\ g_8$. However, all we will attempt in this work, are approximations involving average values.

Assume the momentum vector of $\surd\ (3)\ g_7$ is directed along the + x axis in Cartesian (x,y,z) space. The interacting gluons, $+\ g_8$ and $-\ g_8$, are directed isotropically. Therefore, the average x-component of momentum for both $+\ g_8$ and $-\ g_8$ is zero. When $+\ g_8$ and $-\ g_8$ interact with $\surd\ (3)\ g_7$, the momenta are shared among the five components depicted in Fig. X.3(a), $2R\bar{R}\ /\ \surd\ (2)$, $-\ B\bar{B}\ /\ \surd\ (2)$, $-\ G\bar{G}\ /\ \surd\ (2)$, $+\ B\bar{B}\ /\ \surd\ (2)$, and $-\ G\bar{G}\ /\ \surd\ (2)$, and the five components depicted in Fig. X.3(b), $2R\bar{R}\ /\ \surd\ (2)$, $-\ B\bar{B}\ /\ \surd\ (2)$, $-\ G\bar{G}\ /\ \surd\ (2)$, $-\ B\bar{B}\ /\ \surd\ (2)$, and $+\ G\bar{G}\ /\ \surd\ (2)$.

Recall for {6g1}, the kinetic energies of each of the U_R and D_L quarks are equal. Since we have one U_R quark and two D_L quarks, there is twice the kinetic energy contained in the two D_L as in the single U_R , i.e. $E_{KU} = E_{KD} / 2$. For Fig. X.3(a) and Fig. X.3(b), the situation is reversed, i.e. $E_{KU} = 2 E_{KD}$. This is easily seen, since in Figs. X.3(a) and (b), there are 4 D_L , 2 D_L from $\sqrt{(3)}\, g_7$ and 2 D_L from either $+ g_8$ or $- g_8$. Recalling that the 2 D_L in $\sqrt{(3)}\, g_7$ initially have x-velocities $| v_D | = |2 v_U |$ and the 2 D_L in either $+ g_8$ or $- g_8$ initially have x-velocities $| v_D | = 0$, the 4 D_L in $\sqrt{(3)}\, g_7$ and $\pm g_8$ have shared velocity $| v_D | = | v_U |$.

When the 5-gluon ensembles, Figs. X.3(a) and (b), collapse into the 3-gluon ensembles, Figs. X.4(a) and (b), we have for the U_R and D_L kinetic energies:

$$E_{KU} = m_U (v_U^2/2) = 4\, m_D (v_D^2/2) = 2\, m_D v_D^2$$
$$E_{KD} = 2\, m_D (v_D^2/2) = m_D v_D^2, \text{ then}$$
$$E_{KU} = 2\, E_{KD} .$$

Shown in Fig. X.6 are the normalized energies of the heavy quarks produced by the {6g1} and {6g2} processes, depicting the energy distributions of the U_R and D_L quarks.

Following the appearance of the U_R and D_L quarks and antiquarks produced by the {6g1} and {6g2} processes, we postulate that the quarks and antiquarks combine to form the 4 new, electrically neutral, and lower energy composite states, shown below:

$$\{U_R , D_L , D_L \}^*, \{ \bar{U}_R , \bar{D}_L , \bar{D}_L \}^*, \{ U_R , D_L , D_L \}, \{ \bar{U}_R , \bar{D}_L , \bar{D}_L \}$$

Since these states are structural analogs to neutrons, I have termed them "hyperneutrons," with designations N_1 , N_2 , N_3 , N_4 , respectively. N_1 and N_2 result from the {6g1} process, and N_3 and N_4 result from the

{6g2} process. The asterisks on N_1 and N_2 distinguish the {6g1} hyperneutrons from the {6g2} hyperneutrons, N_3 and N_4 .

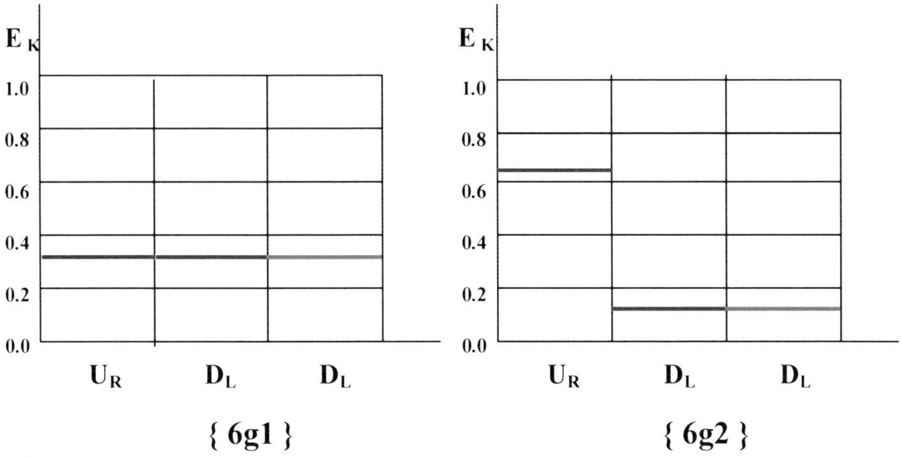

Fig. X.6
Normalized kinetic energy distributions for the heavy quarks produced by the {6g1} and {6g2} processes.

Shown in Fig. X.7 are the kinematic configurations of the U_R and D_L quarks resulting from the {6g1} and {6g2} processes, that lead, in turn, to the four hyperneutrons N_1 , N_2 , N_3 , and N_4 .

We now address the stability of the N_1 , N_2 , N_3 , and N_4 hyperneutrons. In particular, we ask whether or not the hyperneutrons undergo a decay process analogous to the beta decay of ordinary neutrons, i.e. are the composite particles subject to flavor changes via weak decay? Weak decays result in (at a minimum) emissions of W+ or W – bosons, which result in a net change in charge of ± e in the up or down quark resulting from the decay.

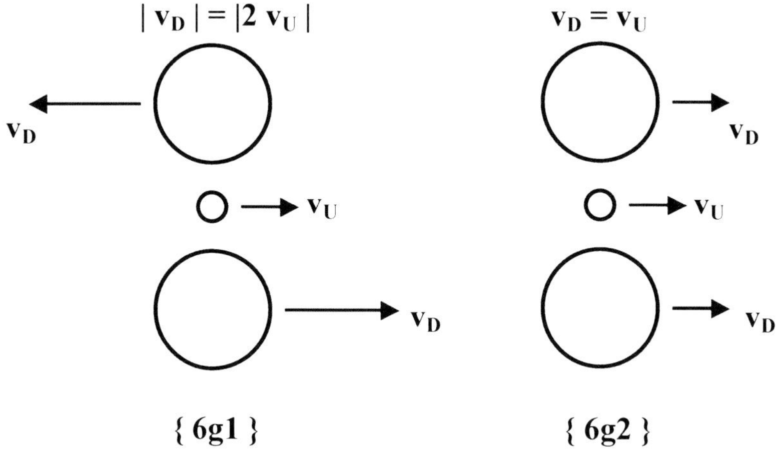

Fig. X.7
Kinematic configurations resulting from the {6g1} and {6g2} processes. v_U and v_D are the velocity vectors of U_R and D_L , respectively.

We postulate that the kinetic energy distributions for the heavy quarks, shown in Fig. X.6 and the kinematic configurations shown in Fig. X.7, determine the stability or instability of the {6g1} or {6g2} hyperneutrons composed of the U_R and D_L heavy quarks. A simple classical analysis was performed for the {6g1} and {6g2} hyperneutrons, comparing center-of-mass (CM) and orbital (O) velocities. For {6g1} hyperneutrons, normalized CM and O velocities were $(2)(6^{-1/2}) / 3 = 0.272$ and $(2)(6^{-1/2}) = 0.816$, respectively. In marked contrast, for {6g2} hyperneutrons, normalized CM and O velocities were $3^{-1/2} = 0.577$ and 0, respectively. These results provided strong justification for postulating that the {6g1} hyperneutrons, N_1 and N_2, are unstable, due to their high O velocities, and subject to weak decay, whereas the {6g2} hyperneutrons, N_3 and N_4 , are stable, due to their zero O velocities.

At this juncture, we postulate that the unstable hyperneutrons N_1 and N_2 are the origins of Regular Matter (RM) and the stable hyperneutrons N_3 and N_4 are the origins of Dark Matter (DM). Since N_1 and N_2 result from the {6g1} process and N_3 and N_4 result from the {6g2} process, we can obtain an approximation for the ratio RM / DM from the ratio of the "footprint" areas shown in Figs. X.1 and X.3. We compare this approximation for RM / DM to the measured value.

The measured value for RM / DM is obtained from [1], where we find the following:

Physical Baryon Density: 0.02260
Physical Dark Matter Density: 0.1123
RM / DM = 0.02260 / 0.1123 = 0.20125

Compare the measured value of RM / DM = 0.201 , to the value of RM / DM calculated from the footprints shown in Figs. X.1, X.3, and X.4 :

From Fig. X.1, we have:

$$A_1 = 16 + 4 + \pi\,(2)^2 = 20 + 4\,\pi$$

Since Fig. X.3 collapses into Fig. X.4, we have:

$$A_2 = 2\,[\sqrt{(3)}]^2\,A_1\text{ , then}$$

$$RM / DM = A_1 / A_2 = A_1 / 6A_1 = 0.167,$$

17% lower than the measured value of 0.201.

Reference

[1] N. Jarosik, et.al., *Seven-Year Wilkinson Microwave Anisotropy Probe (WMAP) Observations: Sky Maps, Systematic Errors,and Basic Results,* Princeton University and NASA Goddard Space Flight Center, ca. 2010, p. 39.

XI The Big-Bang and Hyperneutrons – Part 2

Synopsis

A discussion of weak decays was carried out in [1] and will be reprised here. Assuming the {6g1} hyperneutrons are subject to weak decay, a more detailed look at the decay process involves two virtual particles designated by the notation ( )*. The D and U quark decay processes are as follows:

$$D^{-1/3} \rightarrow (W^{-1} + D^{-1/3})^* \rightarrow U^{+2/3} + l^- + \bar{\nu}_l \qquad \text{(XI. 1)}$$

The U quark decay process is as follows:

$$U^{+2/3} \rightarrow (W^{+1} + U^{+2/3})^* \rightarrow D^{-1/3} + l^+ + \nu_l \qquad \text{(XI. 2)},$$

where l is a lepton and ν_l is a lepton neutrino. Within each of the two virtual particles, the D or U quark carries 1/10 of the total energy of the virtual particle, with 9/10 of the total energy carried by the W bosons. We examine this energy partition and the decay process in the following discussion.

Analysis

A more detailed examination of Eqs. XI.1 and XI.2 lead to a postulated explanation of energy partition between D or U quarks and the W bosons. Eqs. XI.1 and XI.2 are examples of flavor changes, in which a D quark is transformed into a W^{-1} boson and a U quark, or a U quark is transformed into a W^{+1} boson and a D quark.

Two virtual processes are postulated that appear to account for the energy partition that results when a D or U quark is transformed into a W boson and a U or D quark. The most common flavor change is β-decay, where the d quark in a neutron is transformed into a u quark, an electron, and an electron antineutrino.

For the transformation depicted in Eq. XI.1:
$D^{-1/3} \rightarrow (W^{-1} + D^{-1/3})^{*} \rightarrow U^{+2/3} + l^{-} + \bar{\nu}_{l}$,
we postulate the following virtual process, where real particles are written in regular boldface, and virtual particles are written in italics. Ensembles of particles, both real and virtual, are enclosed in braces:

$$\mathbf{D}^{-1/3} \rightarrow \mathbf{D}^{-1/3} + \{U,U,\bar{U},\bar{U},\bar{D}\} + \{\bar{U},D\} \rightarrow \{\mathbf{D},U,U\}^{+1} + \{\bar{D},\bar{U},\bar{U}\}^{-1} + \{\bar{U},D\}^{-1} \rightarrow \{\mathbf{D},U,\bar{U}\}^{-1/3} + \{\bar{D},\bar{U},U\}^{+1/3} + \{\bar{U},D\}^{-1} \rightarrow \mathbf{U}^{+2/3} + \{W\}^{-1} \rightarrow \mathbf{U}^{+2/3} + \mathbf{l}^{-} + \bar{\boldsymbol{\nu}}_{l} \quad \text{(XI. 3)}$$

The dipoles $\{\mathbf{D},U,U\}^{+1}$, $\{\bar{D},\bar{U},\bar{U}\}^{-1}$ and $\{\mathbf{D},U,\bar{U}\}^{-1/3}$, $\{\bar{D},\bar{U},U\}^{+1/3}$ are postulated to be the key ingredients of the flavor change process.

The five steps shown in Eq. XI.3 are as follows:

(1) $\mathbf{D}^{-1/3}$ is joined by seven virtual particles, 2 *U*s, 1 *D*, 3 $\bar{U}$s, and 1 $\bar{D}$.

(2) The energy from $\mathbf{D}^{-1/3}$ is partitioned between the first dipole $\{\mathbf{D},U,U\}^{+1}$, $\{\bar{D},\bar{U},\bar{U}\}^{-1}$ and the virtual meson $\{\bar{U},D\}^{-1}$.
The dipole energy is proportional to the dipole charge squared, $Q^2 = (\pm 1)^2 = 1$.

(3) The dipole $\{\mathbf{D},U,U\}^{+1}$, $\{\bar{D},\bar{U},\bar{U}\}^{-1}$ is transformed into the second dipole $\{\mathbf{D},U,\bar{U}\}^{-1/3}$, $\{\bar{D},\bar{U},U\}^{+1/3}$. The second dipole energy is proportional to the dipole charge squared $Q^2 = (\pm 1/3)^2 = 1/9$. The energy decrement from dipole 1 to dipole 2 is transferred to the virtual meson $\{\bar{U},D\}^{-1}$, which is transformed into the virtual boson $\{W\}^{-1}$.

(4) The energy remaining in dipole 2 is postulated to be transferred to the quark $\mathbf{U}^{+2/3}$.

(5) When $\{W\}^{-1}$ is transformed into l^- and $\bar{\nu}_l$, the process is complete, leaving $\mathbf{U}^{+2/3} + l^- + \bar{\nu}_l$.

For the transformation depicted in Eq. XI.2:
$U^{+2/3} \rightarrow (W^{+1} + U^{+2/3})^* \rightarrow D^{-1/3} + l^+ + \nu_l$,
we postulate a virtual process similar to that used to explain Eq. XI.1:

$$\mathbf{U}^{+2/3} \rightarrow \mathbf{U}^{+2/3} + \{U,D,\bar{U},\bar{U},\bar{D}\} + \{U,\bar{D}\} \rightarrow \{D,\mathbf{U},U\}^{+1} + \{\bar{D},\bar{U},\bar{U}\}^{-1} + \{U,\bar{D}\}^{+1} \rightarrow \{D,\mathbf{U},\bar{U}\}^{-1/3} + \{\bar{D},\bar{U},U\}^{+1/3} + \{U,\bar{D}\}^{+1} \rightarrow \mathbf{D}^{-1/3} + \{W\}^{+1} \rightarrow \mathbf{D}^{-1/3} + \mathbf{l}^+ + \boldsymbol{\nu}_l \qquad \text{(XI. 4)}$$

The five steps shown in Eq. XI.4 are similar to those for Eq. XI.3:

(1) $\mathbf{U}^{+2/3}$ is joined by seven virtual particles, 2 Us, 1 D, 2 $\bar{U}$s, and 2 $\bar{D}$s.

(2) The energy from $\mathbf{U}^{+2/3}$ is partitioned between the first dipole $\{D,\mathbf{U},U\}^{+1}$, $\{\bar{D},\bar{U},\bar{U}\}^{-1}$ and the virtual meson $\{U,\bar{D}\}^{+1}$.

(3) This step is essentially the same as step 3 for Eq. XI.3, with the exception that the virtual meson is $\{U,\bar{D}\}^{+1}$, and the virtual boson is $\{W\}^{+1}$.

(4) The energy remaining in dipole 2 is postulated to be transferred to the quark $\mathbf{D}^{-1/3}$.

(5) When $\{W\}^{+1}$ is transformed into l^- and ν_l, the process is complete, leaving $\mathbf{D}^{-1/3} + l^+ + \nu_l$.

Therefore, the energies of the W bosons are proportional to the square of the charges of the first dipole: $Q^2 = (\pm 1)^2 = 1$, and the energies of the quarks $U^{+2/3}$ and $D^{-1/3}$ are proportional to the square of the charges of the second dipole: $Q^2 = (\pm 1/3)^2 = 1/9$. Therefore, the slope of the quark mass line discussed in Chapter II, is the ratio of the quark energy to the total energy (boson + quark).
$E_q / (E_W + E_q) = (1/9)/(1+ 1/9) = (1/9)/(10/9) = 1/10$.
This is quite close to the observed value of the slope of the quark mass line.

A proposed extended quark sequence was discussed at length in [1]. Following the first weak decay of the heaviest D quark, further downward progression along the sequence line is governed by flavor transformation rules described by a generalization (from 3 x 3 to 11 x 11) of the Cabibbo-Kobayashi-Maskawa Quark-Mixing Matrix (CKM matrix) [2]. The mixing rules, defined by the CKM matrix, only allow +2/3 to –1/3 and –1/3 to +2/3 quark flavor transformations. In contrast, the {6g2} hyperneutrons N_3 and N_4 are stable with respect to weak decay and maintain their identity until they undergo annihilation reactions.

At this juncture, we must consider an explanation for the asymmetry observed between matter and antimatter. Because weak CP violations have been observed among the {u, d, s, c, b, t} quarks, and predicted by the standard CKM matrix, it seems reasonable to postulate that weak decay processes, occurring at much higher energies, could produce much stronger CP violations. Then, substantial baryon number increases would occur as the left-handed quarks and right-handed anti-quarks undergo weak decays. The proposed extension of the CKM matrix to 11 x 11, should allow these processes to be addressed.

The development presented above introduced the four hyperneutrons N_1, N_2 , N_3 , N_4 . N_1 and N_3 are matter and N_2 and N_4 are antimatter. However, only N_3 and N_4 are postulated to be stable, with N_1 and N_2 decaying via weak interactions, down to the stable (or metastable) quarks {u, d, s, c, b, t}. N_3 and N_4 may well be excellent candidates for WIMPS, likely a key constituent of dark matter. N_3 and N_4 should be stable (excepting annihilation reactions), interacting only gravitationally. In fact, the geometrical cross sections of N_3 and N_4 are approximately $2.83 \times 10^{-64} m^2$ (2.83×10^{-36} barns), a value so small as to make N_3 and N_4 excellent candidates for WIMPs.

The tiny cross sections for N_3 and N_4 result in infrequent annihilation reactions, but when annihilations occur, they are highly energetic. Photons of the order of 10^{18} GeV are produced. In contrast, decays of N_1 and N_2 are much more frequent and the photons produced are much more numerous and much less energetic. Therefore, the hyperneutron dynamics introduced above, may produce plausible explanations for the observed photon to baryon ratio and ratio of dark matter-energy to ordinary matter-energy.

As noted in Chapter X, the {6g1} hyperneutrons, N_1 and N_2 , contain two left-handed down quarks (D_L) with significant orbital velocities, and consequently, a propensity for weak decay.

Following the weak decay processes, mediated by plus and minus W Bosons, D_L is converted to U_L , which, in turn, is converted to a less energetic down quark, and so on, as treated at length in the discussion above. The important point is that all of the quarks, deriving initially from D_L , remain left-handed. We postulate that this persistence of left-handedness is responsible for the predominance of left-handed electrons observed in beta decay.

The ordinary quarks, along the quark sequence line, continue to decay via W boson emission until the b quark appears. Since the b quark energy is well below the W boson energy of 80.4 GeV, the decay processes continue via virtual W bosons, until the final decay of the d quark to a u quark. The u quark is essentially stable, because even with virtual W bosons, the decay barrier is too high for further decays to occur.

I have proposed a pair-production scenario in which a primordial gluon ensemble undergoes a transformation into heavy quarks, resulting in both stable and unstable quark complexes. The mixture of stable and unstable quark complexes leads to a new, coherent, and parsimonious model for the Big-Bang, and reasonable candidates for weakly-interacting massive particles (WIMPs). Finally, I discussed possible reasons for the predominance of left-handed electrons observed following beta decay.

However, the hypotheses introduced above will require detailed QCD calculations and corroborative evidence from cosmological observations. In particular, lattice QCD investigations may prove particularly useful.

References

[1] B. P. Shafer, *Quark Heuristics,* Unpublished Manuscript, Seventh Revision, March 10, 2008.

[2] F. J. Gilman, K. Kleinknecht and B. Renk, *The Cabibbo-Kobayashi-Maskawa Quark-Mixing Matrix, Particle Physics Booklet,* U. S. Department of Energy, Pub, 72, July, 2004, p.175.

XII Summary and Conclusion

I have presented a proposal for a new, coherent, and parsimonious scenario for Big-Bang cosmology. A summary of the development, listed by chapter, concludes this discussion:

(1) Chapters I and II describe the motivation for this work and the early work on the extended quark sequence. In [1] an extended quark sequence was suggested, but no end point was specified, as I did in the present work.

(2) Chapter III discusses Frank Wilczek's book and points of departure between my work and Wilczek's.

(3) The quantum vacuum and its essential role in virtual processes is the subject of Chapter IV. I make a distinction between the vacuum and the void, and argue that the void is manifestly stable. I also argue that energy, like temperature, is a positive definite quantity.

(4) Chapter V discusses the initial conditions for the Big-Bang, which I postulate to be a gluon ensemble with initial energy far exceeding the Planck energy.

(5) Chapters VI – IX discuss the Ockham (or Occam) process that I apply to the production of heavy quarks and leptons. An estimate for the size of the electron is presented.

(6) Chapters X and XI discuss the production of two classes of hyperneutrons, one class responsible for the production of Regular Matter (RM), the other class responsible for the production of Dark Matter (DM). I also suggest a mechanism for the production of left-handed electrons following β-decay.

Acknowledgements

As stated earlier, the present work is a sequel to [2] and [3]. The author wishes to acknowledge the contributions of Dr. John Singleton*# and Dr. Mikkel Johnson*# , both who read the first six versions of [2] carefully and made helpful suggestions for improvements. The author also wishes to acknowledge the contributions of Dr. Peter Herczeg*, Dr. J. C. Ingraham, and Dr. O. P. Judd*# who read and commented on later versions of [2] and [3].

∗ Fellow, American Physical Society
Fellow, Los Alamos National Laboratory

References

[1] C. Quigg, *Gauge Theories of the Strong, Weak, and Electromagnetic Interactions,* Westview Press, 1997, p. 152.

[2] B. P. Shafer, Quark Heuristics, Unpublished Manuscript, Seventh Revision, March 10, 2008.

[3] B. P. Shafer, *Gluon Heuristics at t=0,* Unpublished Manuscript, January 16, 2010.

Appendix – Mesons and Baryons

The Constituent Quark Model (CQM) is used in [1] to describe the internal structures of neutrons and protons. In the CQM, neutrons and protons are composed of a core, consisting of virtual quarks and gluons, surrounded by valence quarks. The valence quarks play roles analogous to the roles played by the valence electrons in an atom. More generally, mesons consist of a virtual core surrounded by the quarks and antiquarks, q and $\bar{q}$. Baryons consist of a virtual core surrounded by three quarks, q, q, and q, where q represents any quark. The properties of mesons and baryons are defined by their valence quarks, although the meson and baryon masses are dominated by their respective core masses.

In the discussion that follows, we propose configurations for the mesons $\Pi^{\pm}$, $K^{\pm}$, and K^{0}, and the baryons n^{0}, p^{+}, and Ω^{-}, consisting of ***real***, rather than virtual, combinations of the lightest quarks u, d, and s (or U, D, and S, in the description below).

For the five mesons (Figs. (a) – (e)) and the three baryons (Figs. (f) – (h)), the valence quarks and antiquarks are "outside the box" and the core quarks and antiquarks are "inside the box". For the S quarks, the mass estimates vary widely from 105 ± 25 MeV / c^2 in [2] to 95 ± 25 MeV / c^2 in [3]. For the S valence quarks in Figs. (a) – (h), we assume a mass of 110 MeV / c^2 . For all the U and D quarks in Figs. (a) – (h) we use the masses 4 and 8 MeV / c^2, respectively. The core masses are dominated by S, $\bar{S}$ pairs, whereas the core energies are shared among the S, $\bar{S}$ pairs and the core gluons. In Figs. (a) – (h), we designate the average S quark mass in the core by $\langle S \rangle$, where the average value is obtained by subtracting the valence quark masses and the U and D core quark masses from the total meson or baryon mass (M) , and dividing by the total number of S quarks in the core. All masses are in MeV / c^2 .

We assume the S, $\bar{S}$ pair oscillates between the quark and gluon states as follows:

$$S, \bar{S} \leftrightarrow g\,\bar{g}\,,$$

where the total pair energy oscillates between the S, $\bar{S}$ dipole and the g $\bar{g}$ gluon energy. If the oscillator frequency is ω, then the normalized dipole energy is $[\sin^2 \omega t]$, with peak value 1 and average value 1/2. The normalized dipole mass is the absolute value of $[\sin \omega t]$, which can be written as $\sqrt{}\,[\sin^2 \omega t]$, with peak value 1 and average value $2/\pi$.

We now draw attention to a distinction between Π mesons (Figs. (a) and (b)) and K mesons (Figs. (c), (d), and (e)). The Π meson cores contain only a single S, $\bar{S}$ pair, whereas the K meson cores contain two S, $\bar{S}$ pairs. These configurational differences may have stability implications. For a single pair, the normalized dipole energy and the normalized dipole mass oscillates between 0 and 1, with average values of 1/2 and $2/\pi$, respectively. However, for two pairs, with a phase difference of $\pi/2$, the normalized dipole energy is: $[\sin^2 (\omega t) + \sin^2 (\omega t + \pi/2)] = 1$, and the normalized mass is: $\sqrt{}[\sin^2 (\omega t)] + \sqrt{}[\sin^2 (\omega t + \pi/2)]$. Therefore, the normalized dipole energy of the two-pair configuration is constant and the normalized mass of the two-pair configuration oscillates between the values 1 and $\sqrt{}\,(2)$, admittedly not constant, but less of a fluctuation than 0 to 1. The decrease in mass and energy fluctuations, from Π to K, may be responsible for the remarkable difference between the lifetimes of the Π^0 and K^0 mesons, 8.4×10^{-17} s and 5.1×10^{-8} s, respectively [4].

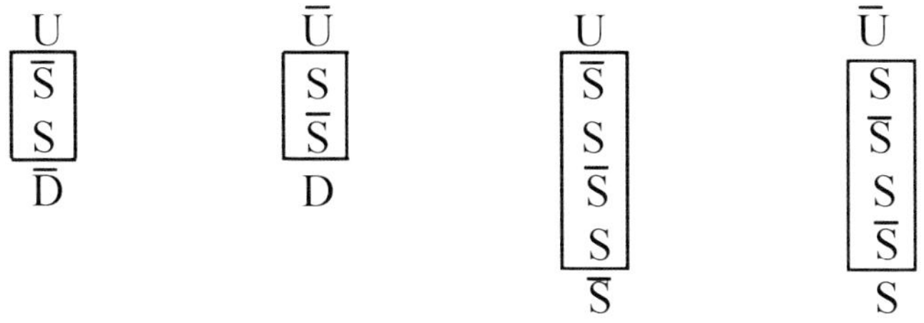

(a) Π^+: M = 140 ⟨S⟩ = 64 (b) Π^-: M = 140 ⟨S⟩ = 64 (c) K^+: M = 494 ⟨S⟩ = 95 (d) K^-: M = 494 ⟨S⟩ = 95

(e) K^0: M = 498
$\langle S \rangle = 95$

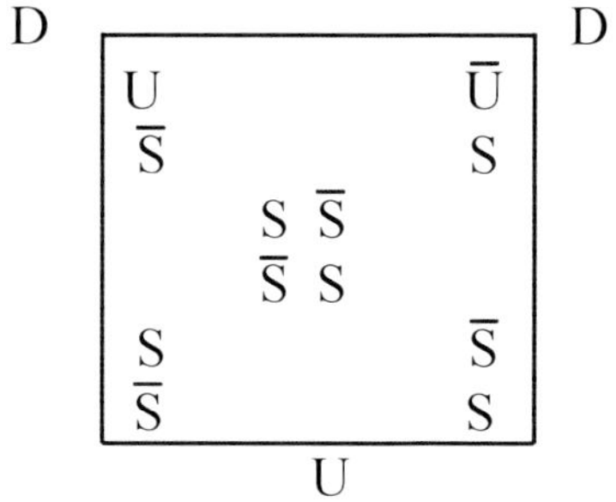

(f) n^0: M = 940
$\langle S \rangle = 91.2$

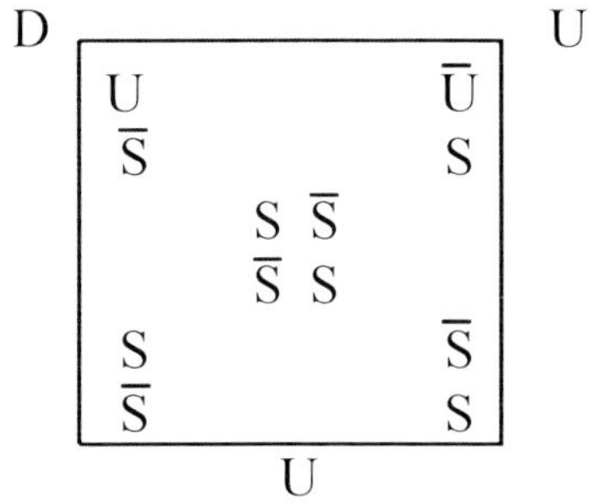

(g) p^+: M = 938
$\langle S \rangle = 91.8$

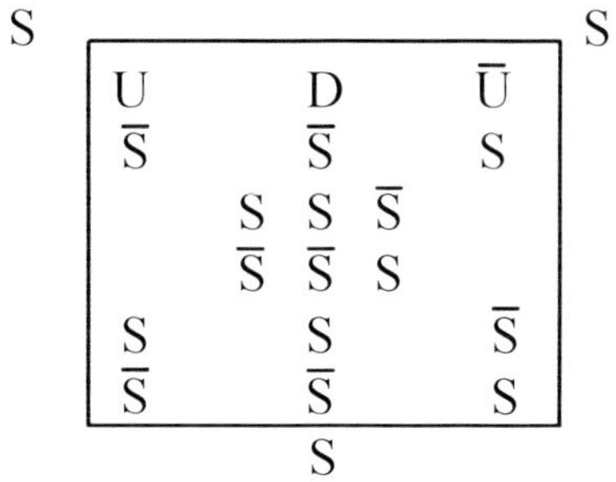

(h) Ω^-: M = 1672
$\langle S \rangle = 88.4$

References

[1] T. P. Smith, *The Anatomy of a Neutron*, American Scientist, Vol. 98, No. 6, November – December, 2010, p. 481.

[2] Particle Data Group, *Particle Physics Booklet,* U. S. Department of Energy, Pub, 72, July, 2004, p. 25.

[3] Particle Data Group, *Particle Physics Booklet,* U. S. Department of Energy, LBL-61033, July, 2006, p. 24.

[4] Ibid. pp. 26, 48.

Sun Books
Sun Publishing

Supplement B-5

MOTIVATIONAL BOOKS FROM THESE FINE AUTHORS

- **James Allen**
- **Lily L. Allen (Mrs. James Allen)**
- **Christian D. Larson**
- **Orison Swett Marden**
- **Ralph Waldo Trine**

JAMES ALLEN

ABOVE LIFE'S TURMOIL by James Allen. True Happiness, The Immortal Man, The Overcoming of Self, The Uses of Temptation, The Man of Integrity, Discrimination, Belief, The Basis of Action, The Belief that Saves, Thought and Action, Your Mental Attitude, Sowing and Reaping, The Reign of Law, The Supreme Justice, The Use of Reason, Self-Discipline, Resolution, The Glorious Conquest, Contentment in Activity, The Temple of Brotherhood, Pleasant Pastures of Peace. 163 pgs. 5x8. Pbk. ISBN: 0-89540-203-3.

ALL THESE THINGS ADDED by James Allen. Entering the Kingdom, The Soul's Great Need, The Competitive Laws and the Law of Love, The Finding of a Principle, At Rest in the Kingdom, The Heavenly Life, The Divine Center, The Eternal Now, The "Original Simplicity", The Unfailing Wisdom, The Might of Meekness, The Righteous Man, Perfect Love, Greatness and Goodness, and Heaven in the Heart. 192 pgs. 5x8. Pbk. ISBN: 0-89540-129-0.

AS A MAN THINKETH by James Allen. Thought and Character, Effect of Thought on Circumstances, Effect of Thought on Health and the Body, Thought and Purpose, The Thought-Factor in Achievement, Visions and Ideals, Serenity. 88 pgs. 5x8. Pbk. ISBN: 0-89540-136-3.

BYWAYS OF BLESSEDNESS by James Allen. Right Beginnings, Small Tasks and Duties, Transcending Difficulties and Perplexities, Burden-Dropping, Hidden Sacrifices, Sympathy, Forgiveness, Seeing No Evil,

Abiding Joy, Silentness, Solitude, Standing Alone, Understanding the Simple Laws of Life, Happy Endings. 202 pgs. 5x8. Pbk. ISBN: 0-89540-202-5.

THE DIVINE COMPANION by James Allen. Truth as Awakener, Truth as Protector, Of Discipline and Purification, Of Purity of Heart, The First Prophecy - Called Awakening, The Fifth Prophecy - Called Transition, The Second Exhortation - Concerning Humility, Instruction Concerning the Great Reality, Discourse Concerning The Way of Truth, Self-Restraint, Etc. 152 pgs. 5x8. Pbk. ISBN: 0-89540-329-3.

EIGHT PILLARS OF PROSPERITY by James Allen. Discussion on Energy, Economy, Integrity, Systems, Sympathy, Sincerity, Impartiality, Self-reliance, and the Temple of Prosperity. 233 pgs. 5x8. Pbk. ISBN: 0-89540-201-7.

ENTERING THE KINGDOM by James Allen. The Soul's Great Need, The Competitive Laws and the Laws of Love, The Finding of a Principle, At Rest in the Kingdom, And All Things Added. 82 pgs. 5x8. Pbk. ISBN: 0-89540-226-2.

FOUNDATION STONES TO HAPPINESS AND SUCCESS by James Allen. Right Principles, Sound Methods, True Actions, True Speech, Equal Mindedness, Good Results. 53 pgs. 5x8. Pbk. ISBN: 0-89540-327-7.

FROM PASSION TO PEACE by James Allen. Passion, Aspiration, Temptation, Transmutation, Transcendence, Beatitude, Peace. 64 pgs. 5x8. Pbk. ISBN: 0-89540-077-4.

FROM POVERTY TO POWER by James Allen. Two books in one: The Path to Prosperity including World a Reflex of Mental States, The Way Out of Undesirable Conditions, Silent Power of Thought, Controlling and Directing One's Forces, The Secret of Health, Success, and Power, Etc. and The Way of Peace including Power of Meditation, The Two Masters, Self and Truth, The Acquirement of Spiritual Power, Realization of Selfless Love, Entering into the Infinite, Perfect Peace, Etc. 184 pgs. 5x8. Pbk. ISBN: 0-89540-061-8.

THE HEAVENLY LIFE by James Allen. The Divine Center, The Eternal Now, The "Original Simplicity," The Unfailing Wisdom, The Might of Meekness, The Righteous Man, Perfect Love, Perfect Freedom, Greatness and Goodness, Heaven in the Heart.84 pgs. 5x8. Pbk. ISBN:0-89540-227-0.

THE LIFE TRIUMPHANT by James Allen. Faith and Courage, Manliness and Sincerity, Energy and Power, Self-Control and Happiness, Simplicity and Freedom, Right-Thinking and Repose, Calmness and Resource, Insight and Nobility, Man and the Master, and Knowledge and Victory. 114 pgs. 5x8. Pbk. ISBN: 0-89540-125-8.

LIGHT ON LIFE'S DIFFICULTIES by James Allen. The Light that Leads to Perfect Peace, The Law of Cause and Effect in Human Life, Values - Spiritual and Material, Adherence to Principle, The Sacrifice of The Self, The Management of the Mind, Self-Control, Acts and Their Consequences, The Way of Wisdom, The Blessing and Dignity of Work, Good Manner and Refinement, Diversity of Creeds, War and Peace, The Brotherhood of Man, Life's Sorrows, Life's Change, Etc. 137 pgs. 5x8. Pbk. ISBN: 0-89540-217-3.

MAN: KING OF MIND, BODY AND CIRCUMSTANCE by James Allen. The Inner World of Thoughts, The Outer World of Things, Habit: Its Slavery and Its Freedom, Bodily Conditions, Poverty, Man's Spiritual Dominion, Conquest. 55 pgs. 5x8. Pbk. ISBN: 0-89540-212-2.

THE MASTERY OF DESTINY by James Allen. Deeds, Character, and Destiny, The Science of Self- Control, Cause and Effect in Human Conduct, Training of the Will, Thoroughness, Mind-Building and Life-Building, Cultivation of Concentration, Practice of Meditation, Power of Purpose, Joy of Accomplishment. 120 pgs. 5x8. Pbk. ISBN: 0-89540-209-2.

MEDITATIONS - A YEAR BOOK by James Allen. "James Allen may truly be called the Prophet of Meditation. In an age of strife, hurry, religious controversy, heated arguments, ritual and ceremony, he came with his message of Meditation, calling men away from the din and strife of tongues into the peaceful paths of stillness within their own souls, where 'the Light that lighteth every man that cometh into the world' ever burns steadily and surely for all who will turn their weary eyes from the strife without to the quiet within." Contains two inspiring quotations and a brief commentary for each day of the year. 366 pgs. 5x8. Pbk. ISBN: 0-89540-192-4.

MEN AND SYSTEMS by James Allen. Men and Systems, Work, Wages, and Well-Being, The Survival of the Fittest as Divine Law, Justice in Evil, Justice and Love, Self-Protection - Animal, Human, and Divine, Aviation and the New Consciousness, The New Courage. 149 pgs. 5x8. Pbk. ISBN: 0- 89540-326-9.

MORNING AND EVENING THOUGHTS by James Allen. Contains a separate and brief paragraph for each morning and evening of the month. 71 pgs. 5x8. Pbk. ISBN: 0-89540-137-1.

OUT FROM THE HEART by James Allen. The Heart and the Life, The Nature of Power of Mind, Formation of Habit, Doing and Knowing, First Steps in the Higher Life, Mental Conditions and Their Effects, Exhortation. 54 pgs. 5x8. Pbk. ISBN: 0-89540-228-9.

THE PATH TO PROSPERITY by James Allen. The Lesson of Evil, The World a Reflex of Mental States, The Way Out of Undesirable Conditions,

The Silent Power of Thought: Controlling and Directing One's Forces, The Secret of Health Success and Power, The Secret of Abounding Happiness, The Realization of Prosperity. 88 pgs. 5x8. Pbk. ISBN: 0-89540-403-6.

POEMS OF PEACE, Including The Lyrical Dramatic Poem EOLAUS by James Allen. 38 Poems. 84 pgs. Pbk. ISBN: 0-89540-429-X.

THE SHINING GATEWAY by James Allen. The Shining Gateway of Meditation, Temptation, Regeneration, Actions and Motives, Morality and Religion, Memory, Repetition and Habit, Words and Wisdom, Truth Made Manifest, Spiritual Humility, Spiritual Strength, Etc. 58 pgs. 5x8. Pbk. ISBN: 0-89540-328-5.

THROUGH THE GATE OF GOOD by James Allen. The Gate and the Way, The Law and the Prophets, The Yoke and the Burden, The Word and the Doer, The Vine and the Branches, Salvation this Day. 66 pgs. 5x8. Pbk. ISBN: 0-89540-216-5.

THE WAY OF PEACE by James Allen. The Power of Meditation, The Two Masters: Self and Truth, The Acquirement of Spiritual Power, The Realization of Selfless Love, Entering into the Infinite, Saints, Sages, and Saviors, The Law of Service, The Realization of Perfect Peace. 113 pgs. 5x8. Pbk. ISBN: 0-89540-229-7.

Lily L. Allen (Mrs. James Allen)

THE BRYNGOLEU COOKERY BOOK. A Humanitarian Household Guide by Lily L. Allen. Breakfast Menus for One Week, Home, Luncheon Menus, The Cuisine of the Compassionate, Dinner Menus, Hints to Housewives, Soups, Sauces, Gravies, Sweets, Bread and Cakes, Etc. 66 pgs. 5x8. Pbk. ISBN: 0-89540-456-7.

CLUES TO THE TREASURES OF LIFE by Lily L. Allen. Chapters include: Clues to Life's Treasures, Clues to the Secret of Health, The Greatest of All Clues, Clues to the Abundant Life, Clues to the Glorious Destination of Man, Finding the Deathless Life, The Law of Attraction, The Will of God Concerning Man, Clues for Everyday Life, Etc. 83 pgs. 5x8. Pbk. ISBN: 0-89540-451-6. (ISBN-13: 978-0-89540-451-0).

ELEMENTS OF SUCCESS by Lily L. Allen. The Fine Art of Living, The Essential Ideal, The Survival of the Fittest, Thought - The Greatest Element in Success, Difficulties, Drawbacks, The Equal Mind, The Success Habit, Climbing to Success, Etc. 136 pgs. 5x8. Pbk. ISBN: 0-89540-446-X.

GUIDE TO A HEALTHY, HAPPY, AND HUMANE LIFE by Lily L. Allen. 130 pgs. 5x8. Pbk. ISBN: 0-89540-460-5.

HEALING FOR ALL OUR ILLS by Lily L. Allen. Clearing the Mind, Reversing the Thoughts, Self-Treatment, We Are in Heaven Now, The Use of Denials, The Use of Affirmations, Healing Treatments, Mixed-Thinking, Thinking of Others, The Need of Patience, The Power of Words, Perfect Health, The Abundant Life, The Ideal is Always the Real, The Vision of the Healer, Etc. 79 pgs. 5x8. Pbk. ISBN: 0-89540-453-2.

HOW TO GO INTO THE SILENCE by Lily L. Allen. Chapters include: Going Into the Silence, How to Enter the Silence, Why We Go Into the Silence, Waiting In the Silence. 23 pgs. 5x8. Pbk. ISBN: 0-89540-452-4.

IN THE GARDEN OF SILENCE by Lily L. Allen. Told in the Twilight, Immortality, Opportunity, Penelope, My Gold!, Afterward, Unbroken Alabaster. 59 pgs. 5x8. Pbk. ISBN: 0-89540-420-6.

THE KNOWLEDGE THAT BRINGS PEACE by Lily L. Allen. Chapters include: The Knowledge That Brings Peace, The Law of the Increase of All Good, A New Order of Thought, A New Method of Thinking. 47 pgs. 5x8. Pbk. ISBN: 0-89540-455-9.

THE LIFE OF POWER AND DOMINION by Lily L. Allen. The Outlook, What Man Is, The One Creator, Affirming the Truth, The Law of Growth, The Perfect Harmony, The Outpouring of All Good, Believing and Receiving, The Life of Power and Dominion, Work: The Glory of Life, Inheritors of All Good. 64 pgs. 5x8. Pbk. ISBN: 0-89540-449-4.

LIFE'S INSPIRATIONS by Lily L. Allen. Unseen Forces, Beauty, Nature, Friendship, Smiles, Work, Companionship and Solitude, Sorrow, Thinking to Inspire, What We Call Death, Life's Greatest Inspiration, Etc. 167 pgs. 5x8. Pbk. ISBN: 0-89540-457-5. (ISBN-13: 978-0-89540-457-2).

THE MEANING AND USE OF THE SILENCE by Lily L. Allen. The True Knowledge of Prayer, The Prayer That is Always Answered, The Realisation of Our Sonship, The Soul's Trysting Place and Time, The Silence for Meditation, Why the Silence is Necessary, The Power of the Silence in the Day's Work, Perfect Rest and Confidence Through the Silence. 55 pgs. 5x8. Pbk. ISBN: 0-89540-454-0.

THE MIGHT OF MIND by Lily L. Allen. Control of the Mind, Creative Power of Mind, Thought - The Alchemist, Desire or Aspiration, What Will Ye? The Influence of Thought on Environment, The Philosopher's Stone, "With All Thy Getting." 79 pgs. 5x8. Pbk. ISBN: 0-89540-443-5.

THE MINISTRY OF SILENT HELPERS by Lily L. Allen. The Ministry of Silent Helpers, The Divine Relationship and What It Implies, Seeing As God Sees, The Power of Right Thinking, The Master of My Fate, Seeking First the Kingdom. 87 pgs. 5x8. Pbk. ISBN: 0-89540-448-6.

ONE LIFE, ONE LAW, ONE LOVE by Lily L. Allen. The World Within, Inward Light, Becoming, Law of Love, Limitations, Taking Time to Think, Where Understanding Dwells. 57 pgs. Pbk. ISBN: 0-89540-445-1.

OUR CREATIVE IMAGINATION by Lily L. Allen. Imagination: What Is It?, The Law of Correspondences, The Eternal Now, The Eternal Memory, Mental Impressions, Divine Alchemy, Removing Mountains, The Eternal Good, The Pathway of the Conqueror. 50 pgs. 5x8. Pbk. ISBN: 0-89540-450-8. (ISBN-13: 978-0-89540-450-3).

OUR MENTAL CHILDREN by Lily L. Allen. Our Thought Children, Fear, Pride, Suspicion, Sincerity, Hope, The Beautiful, Some Simple Rules. 42 pgs. Pbk. ISBN: 0-89540-447-8.

PERSONALITY: ITS CULTIVATION AND POWER AND HOW TO ATTAIN by Lily L. Allen. Personality, Right Belief, Self-Knowledge, Intuition, Decision and Promptness, Self-Trust, Thoroughness, Manners, Physical Culture, Mental, Moral, and Spiritual Culture, Introspection, Emancipation, Self-Development, Self-Control and Mental Poise, Liberty, Transformation, Balance, Meditation and Concentration. 170 pgs. 5x8. Pbk. ISBN: 0-89540-218-1.

CHRISTIAN D. LARSON

BRAINS AND HOW TO GET THEM by Christian D. Larson. Building the Brain, Making Every Brain Cell Active, Principles in Brain Building, Practical Methods in Brain Building, Vital Secrets in Brain Building, Special Brain Development, The Inner Secret, The Finer Forces, Subjective Concentration, Principle of Concentration, Development of Business Ability, Accumulation and Increase, Individual Advancement, The Genius of Invention, The Musical Prodigy, Talent and Genius in Art, Talent and Genius in Literature, Vital Essentials In Brain Building. 233 pgs. 5x8. Pbk. ISBN 0-89540-382-X.

BUSINESS INSPIRATIONS by Christian D. Larson. Chapters include: Follow the High Vision, Inspiration Moves the World, This Will Mean Success, How We Gain Power, The Good Cheer Attitude, The Magic of Sincerity, It Pays to Look Well, The Full Positive Action, Virtues of Persistent Desire, Demand More of Yourself, What Makes Men Great, Giving Your Whole Self, Knowing How to Work, The Whole Mind in Action, Creating the Right Idea, The Will to Do More, etc. 163 pgs. 5x8. Pbk. ISBN: 0-89540-399-4.

THE HIDDEN SECRET by Christian D. Larson. There is a life within that has no limit; it is the life more abundant - the life that every awakened mind has sought with heart and soul; it is the life from which all great things proceed; the source of everything that has real value and high worth. 101 pgs. 5x8. Pbk. ISBN: 0-89540-404-4.

HOW GREAT MEN SUCCEED by Christian D. Larson. The man who succeeds is invariably impelled to press on and on by something within him that tells him he can... In the spirit of this conviction he proceeds with his eye singled upon the goal in view, gaining ground steadily and demonstrating every day, through actual results, that his conviction is based upon fact. 95 pgs. 5x8. Pbk. ISBN: 0-89540-398-6.

JUST BE GLAD by Christian D. Larson. For great is the power of sunshine, especially human sunshine. It can change anything, transform anything, remake anything, and cause anything to become as fair and beautiful as itself. 64 Pages. 5x8. Pbk. ISBN: 0-89540-441-9.

MASTERY OF FATE by Christian D. Larson. What man is, and what man does, determines in what conditions, circumstances, and environments he shall be placed. Since man can change both himself and his actions, he can determine what his fate is. 95 pgs. 5x8. Pbk. ISBN: 0-89540-383-8.

MASTERY OF SELF by Christian D. Larson. No power in man can do what it is created to do, and what it has the capacity to do, until it is directed by man himself; powers, elements, forces, and things are at the disposal of man; they can do only what he directs them to do; they respond only to his control, but before man can gain the power to master forces and things, he must gain the power to master himself. 87 pgs. 5x8. Pbk. ISBN: 0-89540-384-6.

NOTHING SUCCEEDS LIKE SUCCESS by Christian D. Larson. When we think of success, we usually think of the accumulation of wealth, but this can never be more than a small fragment of success, because success in reality signifies any form of attainment or achievement that is truly worth while. 80 pgs. 5x8. Pbk. ISBN: 0-89540-385-4.

ON THE HEIGHTS by Christian D. Larson. We are under the clear sky of Infinite Light, on the verge of the great beyond, on the border-land of the limitless, and every moment is an eternity of ecstasy divine. 74 pgs. 5x8. Pbk. ISBN: 0-89540-442-7.

PERFECT HEALTH by Christian D. Larson. To apply the principle of faith in the realization of perfect health, enter into the spirit of perfect health whenever you think of health. Live in the soul of health, or the real, interior life of health. 78 pgs. 5x8. Pbk. ISBN: 0-89540-397-8.

PRACTICAL SELF-HELP by Christian D. Larson. Chapters include: Lean to Help Yourself, What You Should Do, Changing Your Own World, The Successful Mental Attitude, Full Use of Ability and Power, The Control of Circumstance, Effective Use of Thought and Action, Vital Principles in Self-Help, Building Self-Confidence, Invincible Determination, Know What You Want, The Control of Things, There is

Always a Way, Actions That Produce Response, Directing the Forces of Life, The Most Helpful Principle Known Today, Etc! 223 pgs. 5x8. Pbk. ISBN: 0-89540-400-1.

THINKING FOR RESULTS by Christian D. Larson. That man can change himself, improve himself, recreate himself, control his environment, and master his own destiny is the conclusion of every mind that is wide-awake to the power of right thought in constructive action. In fact, it is the conviction of all such minds that man can do practically anything within the possibilities of the human domain when he knows how to think, and that he can secure almost any result desired when he learns how to think for results. 132 pgs. 5x8. Pbk. ISBN: 0-89540-381-1.

WHAT RIGHT THINKING WILL DO by Christian D. Larson. Whenever we think, we form a mental image of that about which we are thinking; or we may speak of it as a mental picture; and in the science of right thinking, the question is, what kind of mental pictures or images we shall form, because we understand the fact that every mental image becomes a pattern in the mind, and the creative energies of the mind will produce states, conditions, and qualities that correspond exactly with the mental image. 50 pgs. 5x8. Pbk. ISBN: 0-89540-386-2.

YOUR FORCES AND HOW TO USE THEM by Christian D. Larson. How We Govern the Forces We Possess, The Use of Mind in Practical Action, Training the Subconscious for Special Results, How Man Becomes What He Thinks, He Can Who Thinks He Can, How We Secure What We Persistently Desire, Concentration and the Power of Suggestion, The Development of the Will, The Building of a Great Mind, How Character Determines Constructive Action, The Creative Forces in Man, Imagination and the Master Mind, Etc. 331 pgs. 6x8. Pbk. ISBN 0-89540-380-3.

ORISON SWETT MARDEN

AMBITION AND SUCCESS by Orison Swett Marden. What is Ambition, The Satisfied Man, The Influence of Environment, Unworthy Ambitions, Ambition Knows No Age Limit, Make Your Life Count, Visualize Yourself in a Better Position, Thwarted Ambition, Why Don't You Begin? 75 pgs. 5x8. Pbk. ISBN: 0-89540-369-2.

BE GOOD TO YOURSELF by Orison Swett Marden. 322 pgs. 5x8. Pbk. ISBN: 0-89540-364-1.

CHARACTER - THE GRANDEST THING IN THE WORLD by Orison Swett Marden. A Grand Character, The Light Bearers, The Great-Hearted, Intrepidity of Spirit, "A Fragment of the Rock of Ages," Etc. 55 pgs. Pbk. ISBN: 0-89540-297-1.

CHEERFULNESS AS A LIFE POWER by Orison Swett Marden. Inspiration, Self-Help, Positive Thinking, Recovery, Business Etiquette. 79 pgs. 5x8. Pbk. ISBN: 0-89540-363-3.

CHOOSING A CAREER by Orison Swett Marden. 332 pgs. 5x8. Pbk. ISBN: 0-89540-412-5.

THE CONQUEST OF WORRY by Orison Swett Marden. 535 pgs. 5x8. Pbk. ISBN: 0-89540-396-X

EVERY MAN A KING, or Might in Mind Mastery by Orison Swett Marden. Steering Thought Prevents Life Wrecks, How Mind Rules the Body, Thought Causes Health and Disease, Overcoming Fear, Mastering our Moods, The Power of Cheerful Thinking, Affirmation Creates Power, How Thinking Brings Success, Building Character, The Power of Imagination, How to Control Thought, Etc. 240 pgs. 5x8. Pbk. ISBN: 0-89540-334-X.

EVERYBODY AHEAD OR GETTING THE MOST OUT OF LIFE by Orison Swett Marden. 535 pgs. 5x8. Pbk. ISBN: 0-89540-409-5.

THE EXCEPTIONAL EMPLOYEE by Orison Swett Marden. Inspiration, positive thinking, self-help, business. 202 pgs. 5x8 pgs. Pbk. ISBN: 0-89540-352-8.

GETTING ON by Orison Swett Marden. 325 pgs. 5x8. Pbk. ISBN: 0-89540-370-6.

GOOD MANNERS - A PASSPORT TO SUCCESS by Orison Swett Marden. 64 pgs. 5x8. Pbk. ISBN:0-89540-366-8.

HE CAN WHO THINKS HE CAN by Orison Swett Marden. He Can Who Thinks He Can, Getting Aroused, Education by Absorption, Freedom at Any Cost, What the World Owes to Dreamers, The Spirit in Which You Work, Responsibility Develops Power, Stand for Something, Happy, If Not, Why? Originality, Sizing Up People, Getting Away From Poverty, Etc. 245 pgs. 5x8. Pbk. ISBN: 0-89540-346-3.

THE HOUR OF OPPORTUNITY by Orison Swett Marden. The Hour of Opportunity: Are You Ready For It? Self-Made or Never Made, Do Not Wait for Opportunity, Self-Training, Do You Know a Good Thing When You See It? Every-Day Opportunities, The Executive Quality, What is My Right Place, "I Never Asked Anything About It," The Power of Adaptation, Focus Your Energies, Become a Specialist, The Inspiration of a Great Purpose, Etc. 72 pgs. 5x8. Pbk. ISBN: 0-89540-336-6.

HOW THEY SUCCEEDED - LIFE STORIES OF SUCCESSFUL MEN AND WOMEN TOLD BY THEMSELVES by Orison Swett Marden. Marshall Field, Alexander G. Bell, Helen Gould, Philip D.

Armour, Mary E. Proctor, John Wanamaker, Darius Ogden Mills, Lillian Nordica, John D. Rockefeller, Julia Ward Howe, Thomas A. Edison, Lew Wallace, Andrew Carnegie, John Burroughs, James Whitcomb Riley, Etc. 365 pgs. 5x8. Pbk. ISBN: 0-89540-345-5.

HOW TO SUCCEED or STEPPING STONES TO FAME AND FORTUNE by Orison Swett Marden. 332 pgs. 5x8. Pbk. ISBN: 0-89540-371-4.

HOW TO GET WHAT YOU WANT by Orison Swett Marden. How to Get What You Want, Discouragement a Disease, How to Cure It, The Force that Moves Mountains, Faith and Drugs, How to Find Oneself, How to Attract Prosperity, Heart-to-Heart Talks With Yourself, Etc. 331 pgs. 5x8. Pbk. ISBN: 0-89540-335-8.

AN IRON WILL by Orison Swett Marden. Training the Will, Mental Discipline, Conscious Power, Do You Believe in Yourself? Will Power in its Relation to Health and Disease, The Romance of Achievement Under Difficulties, Concentrated Energy, Staying Power, Persistent Purpose, Success Against Odds. 52 pgs. Pbk. ISBN: 0-89540-283-1.

THE JOYS OF LIVING OR LIVING TODAY IN THE HERE AND NOW by Orison Swett Marden. 403 pgs. 5x8. Pbk. ISBN: 0-89540-389-7.

LITTLE VISITS WITH GREAT AMERICANS. TWO VOLUME SET by Orison Swett Marden. 742 pgs. 5x8. Pbk. ISBN: 0-89540-374-9.

LITTLE VISITS WITH GREAT AMERICANS. VOL. II by Orison Swett Marden. 389 pgs. 5x8. Pbk. ISBN: 0-89540-373-0.

LITTLE VISITS WITH GREAT AMERICANS. VOL. I by Orison Swett Marden. 352 pgs. 5x8. Pbk. ISBN: 0-89540-372-2.

MAKING LIFE A MASTERPIECE by Orison Swett Marden. 329 pgs. 5x8. Pbk. ISBN: 0-89540-365-X.

THE MAKING OF A MAN by Orison Swett Marden. 307 pgs. 5x8. Pbk. ISBN: 0-89540-408-7.

MAKING YOURSELF by Orison Swett Marden. 320 pgs. 5x8. Pbk. ISBN: 0-89540-413-3.

THE MIRACLE OF RIGHT THOUGHT by Orison Swett Marden. Working for One Thing and Expecting Something Else, Expect Great Things of Yourself, Self-Encouragement by Self-Suggestion, Change the Thought - Change the Man, The Paralysis of Fear, Getting in Tune, A New Way of Bringing Up Children, Training for Longevity, As A Man Thinketh, Etc. 339 pgs. Pbk. ISBN: 0-89540-311-0.

NOT THE SALARY BUT THE OPPORTUNITY by Orison Swett Marden. 96 pgs. 5x8. Pbk. ISBN: 0-89540-410-9.

THE OPTIMISTIC LIFE by Orison Swett Marden. The Power of Amiability, The Inner Life as Related to Outward Beauty, The Value of Friends, The Cost of an Explosive Temper, Learn to Expect a Great Deal of Life, Mental Power, If You Can Talk Well, Brevity and Directness, What Distinguishes Work From Drudgery, Keeping Fit for Work, Mastering Moods, Business Integrity, Wresting Triumphant from Defeat, Freshness in Work, Don't Take Your Business Troubles Home, Let It Go, Etc! 316 pgs. 5x8. Pbk. ISBN: 0-89540-351-X.

PEACE, POWER, AND PLENTY by Orison Swett Marden. The Power of the Mind to Compel the Body, Poverty a Mental Disease, The Law of Opulence, Character-Building and Health-Building During Sleep, Health Through Right Thinking, Imagination and Health, How Suggestion Influences Health, Why Grow Old?, The Miracle of Self-Confidence, Self-Control vs the Explosive Passions, Good Cheer - God's Medicine, Etc. 323 pgs. 5x8. Pbk. ISBN: 0-89540-343-9.

THE POWER OF PERSONALITY by Orison Swett Marden. Inspiration, Self-Help, Positive Thinking, Recovery, Business Etiquette. 86 pgs. 5x8. Pbk. ISBN: 0-89540-362-5.

THE PROGRESSIVE BUSINESS MAN or How the Right Mental Attitude and Reciprocity are Revolutionizing Business by Orison Swett Marden. 166 pgs. 5x8. Pbk. ISBN: 0-89540-390-0.

PROSPERITY, HOW TO ATTRACT IT by Orison Swett Marden. 325 pgs. 5x8. Pbk. ISBN: 0-89540-392-7.

PUSHING TO THE FRONT VOL I by Orison Swett Marden. Opportunities Where You Are, Possibilities in Spare Time, How Poor Boys and Girls Go to College, Your Opportunity Confronts You, What Will You Do With It? Choosing a Vocation, Concentrated Energy, Triumph of Enthusiasm, Promptness, Appearance, Personality, Common Sense, Accuracy, Persistence, Success Under Difficulties, Observation and Self-Improvement, Triumph of the Common Virtues. 432 pgs. 5x8. Pbk. ISBN: 0-89540-331-5.

PUSHING TO THE FRONT VOL II by Orison Swett Marden. The Man With an Idea, The Will and the Way, Work and Wait, Might of Little Things, Expect Great Things of Yourself, Stand for Something, Habit: Servant or Master, Power of Purity, Power of Suggestion, Conquest of Poverty, Home as a School of Good Manners, Thrift, Why Some Succeed and Others Fail, Character is Power, Rich Without Money. 441 pgs. 5x8. Pbk. ISBN: 0-89540-332-3.

PUSHING TO THE FRONT - TWO VOL. SET by Orison Swett Marden. 873 pgs. ISBN: 0-89540-333-1.

RISING IN THE WORLD OR ARCHITECTS OF FATE by Orison Swett Marden. 318 pgs. 5x8. Pbk. ISBN: 0-89540-375-7.

THE SECRET OF ACHIEVEMENT by Orison Swett Marden. Moral Sunshine, "Blessed Be Drudgery", Honesty - As Principle and As Policy, Habit: The Servant or The Master, Courage, Self-Control, & the School of Life, Decide, Tenacity of Purpose, The Art of Keeping Well, Purity is Power, Etc. 301 pgs. 5x8. Pbk. ISBN: 0-89540-337-4.

SELLING THINGS by Orison Swett Marden. The Man Who Can Sell Things, Training the Salesman, Making a Favorable Impression, The Selling Talk or "Presentation", How to Get Attention, Friend - Winner and Business-Getter, Sizing Up the Prospect, How Suggestion Helps in Selling, The Gentle Art of Persuasion, Closing the Deal, Enthusiasm, Meeting and Forestalling Objections, Finding Customers, When You are Discouraged, Know Your Goods, Character is Capital, Keeping Fit and Salesmanship, Etc. 276 pgs. 5x8. Pbk. ISBN: 0-89540-339-0.

STORIES FROM LIFE A BOOK FOR YOUNG PEOPLE by Orison Swett Marden. 240 pgs. 5x8. Pbk. ISBN: 0-89540-411-7.

SUCCESS NUGGETS by Orison Swett Marden. Inspiration, Positive Thinking, Self-help, Business. 76 pgs. 5x8. Pbk. ISBN: 0-89540-354-4.

TALKS WITH GREAT WORKERS by Orison Swett Marden. 110 pgs. 5x8. Pbk. ISBN: 0-89540-402-8.

THRIFT by Orison Swett Marden. 92 pgs.5x8. Pbk. ISBN:0-89540-393-5.

TRAINING FOR EFFICIENCY by Orison Swett Marden. 360 pgs. 5x8. Pbk. ISBN: 0-89540-394-3.

THE VICTORIOUS ATTITUDE by Orison Swett Marden. Inspiration, Positive Thinking, Self-Help, Business. 358 pgs. 5x8. Pbk. ISBN: 0-89540-353-6.

WHY GROW OLD? by Orison Swett Marden. Marden instructs his reader to "hold to youthful, buoyant thought" and keep the imagination alive and flexible. Recognizing that we may be slaves to our attitudes, this text encourages us to make as much of ourselves as possible and in doing so watch as our lives are prolonged. 30 pgs. 5x8. Pbk. ISBN: 0-89540-340-4.

WINNING OUT, A Book for Young People on Character Building by Habit Forming by Orison Swett Marden. 251 pgs. 5x8. Pbk. ISBN: 0-89540-377-3.

YOU CAN, BUT WILL YOU? by Orison Swett Marden. The Magic Mirror, The New Philosophy of Life, Connecting With the Power that Creates, You Can, But Will You? How Do You Stand With Yourself? The New Philosophy in Business, What Are You Thinking? Facing Life the Right Way, How to Realize Your Ambition, The Open Door, Do You Carry Victory in Your Face? 338 pgs. 5x8. Pbk. ISBN: 0-89540-342-0.

THE YOUNG MAN ENTERING BUSINESS by Orison Swett Marden. 307 pgs. 5x8. Pbk. ISBN: 0-89540-378-1.

RALPH WALDO TRINE

CHARACTER BUILDING THOUGHT POWER by Ralph Waldo Trine. "Have we within our power to determine at all times what types of habits shall take form in our lives? In other words, is habit-forming, character-building, a matter of mere chance, or do we have it within our control?" 51 pgs. 5x8. Pbk. ISBN 0-89540-251-3.

EVERY LIVING CREATURE or Heart Training Through the Animal World, by Ralph Waldo Trine. The tender and humane passion in the human heart is too precious a quality to allow it to be hardened or effaced by practices such as we often indulge in. 50 pgs. 5x8. Pbk. ISBN 0-89540-309-9.

THE GREATEST THING EVER KNOWN by Ralph Waldo Trine. The Greatest Thing Ever Known, Divine Energies in Every-Day Life, The Master's Great but Lost Gift, The Philosopher's Ripest Life Thought, Sustained in Peace and Safety Forever. 57 pgs. 5x8. Pbk. ISBN 0-89540-274-2.

THE HIGHER POWERS OF MIND AND SPIRIT by Ralph Waldo Trine. The Silent, Subtle Building Forces of Mind and Spirit, Thought as a Force in Daily Living, The Divine Rule in the Mind and Heart, The Powerful Aid of the Mind In Rebuilding Body-How Body Helps Mind, Etc. 240 pgs. 5x8. Pbk. ISBN 0-89540-278-5.

IN THE FIRE OF THE HEART by Ralph Waldo Trine. With the People: A Revelation, The Conditions that Hold Among Us, As Time Deals with Nations, As to Government, A Great People's Movement, Public Utilities for the Public Good, Labour and Its Uniting Power, Agencies Whereby We Shall Secure the People's Greatest Good, The Great Nation, The Life of the Higher Beauty and Power. 338 pgs. 6x8 Pbk. ISBN 0-89540-310-2.

IN THE HOLLOW OF HIS HAND by Ralph Waldo Trine. The Present Demand to Know the Truth, The Thought - The Existing Conditions and the Religions of Jesus' Time, What Jesus Realized, Jesus' Own Statement of the Essence of Religion, Was the Church Sanctioned or Established by

Jesus?, Our Debt to the Prophets of Israel, The Power - The Beauty - and the Sustaining Peace. 242 pgs. 5x8. Pbk. ISBN 0-89540-358-7.

IN TUNE WITH THE INFINITE by Ralph Waldo Trine. A Message to My Readers, Prelude, The Supreme Fact of the Universe, The Supreme Fact of Human Life, Fullness of Life - Bodily Health and Vigor, The Secret Power and Effects of Love, Wisdom and Interior Illumination, The Realization of Perfect peace, Coming into Fullness of Power, Plenty of All Things - The Law of Prosperity, How Men Have Become Prophets, Seers, Sages, and Saviors, The Basic Principle of All Religions - The Universal Religion, Entering Now Into the Realization of the Highest Riches, Etc. 221 pgs. 5x8. Pbk. ISBN: 0-89540-387-0.

THE MAN WHO KNEW by Ralph Waldo Trine. The Power of Love, All is Well, That Superb Teaching of "Sin," He Teaches the Great Truth, When a Brave Man Chooses Death, Bigotry in Fear Condemns and Kills, Love the Law of Life, The Creative Power of Faith and Courage, Etc. 230 pgs. 5x8. Pbk. ISBN 0-89540-267-X.

MY PHILOSOPHY AND MY RELIGION by Ralph Waldo Trine. This Place: Amid the Silence of the Centuries, With the Oldest Living Things, My Philosophy, My Religion, The Creed of the Open Road. 130 pgs. 5x8. Pbk. ISBN 0.89540-349-8.

THE NEW ALIGNMENT OF LIFE by Ralph Waldo Trifle. Science and Modern Research, The Modern Spiritual Revival, The Vitalizing Power of the Master's Message, Modern Philosophic Thought, A Thinking Man's Religion, A Healthy Mind in a Healthy Body, The Mental Law of Habit. 228 pgs. 5x8. Pbk. ISBN 0-89540-347-1.

ON THE OPEN ROAD by Ralph Waldo Trine. To always clearly realize that thoughts are forces, that like creates like and like attracts like, and that to determine one's thinking therefore is to determine his life. 65 pgs. 5x8. Pbk. ISBN 0-89540-252-1.

THIS MYSTICAL LIFE OF OURS. A Book of Suggestive Thoughts for Each Week Through the Year by Ralph Waldo Trine. The Creative Power of Thought, The Laws of Attraction, Prosperity, and Habit-Forming, Faith and Prayer - Their Nature, Self-Mastery, Thoughts are Forces, How We Attract Success or Failure, The Secret and Power of Love, Will - The Human and The Divine, The Secret of the Highest Power, Wisdom or Interior Illumination, How Mind Builds Body, Intuition: The Voice of the Soul, To Be at Peace, Etc. 190 pgs. 5x8. Pbk. ISBN: 0-89540-279-3.

THROUGH THE SUNLIT YEAR by Ralph Waldo Trine. A book of Suggestive Thoughts for each day of the year from the writings of Ralph Waldo Trine. 250 pgs. 6x8. Pbk. ISBN: 0-89540-350-1.

WHAT ALL THE WORLD'S A-SEEKING, or the Vital Law of True Life, True Greatness, Power and Happiness by Ralph Waldo Trine. The Principle, Application, Unfoldment, Awakening, The Incoming, Character-Building Thought Power. 224 pgs. 5x8. Pbk. ISBN: 0-89540-359-5.

THE WINNING OF THE BEST by Ralph Waldo Trine. Which Way Is Life Leaning? The Creative Power of Thought, The Best Is the Life, The Power That Makes Us What We Are, A Basis of Philosophy and Religion, How We Will Win the Best. 100 pgs. 5x8. Pbk. ISBN: 0-89540-348-X.

ADDITIONAL TITLES

ACRES OF DIAMONDS by Russell H. Conwell. Through this lecture, Conwell debunks the idea that it is noble to be poor, an idea that far too many of us share. He illustrates that it is our duty as humans to use our gifts to honestly earn riches, because we can do more good with riches than without. 82 pgs. 5x8. Pbk. ISBN: 0-89540-418-4.

BEING AND BECOMING - The Principles and Practices of the Science of Spirit by Fenwicke L. Holmes. The Great Law of Mind, Concentration vs. Ideation, Affirmation, Healing Realization, The Purpose of Spirit, The Motive - Love, Love - The Healing Power, Feelings and Emotions, Why Many Fail, Mysticism, Our Power of Choice, Being, Intuition, Spirit as Formative, Demonstrating Prosperity. 50 pgs. 5x8. Pbk. ISBN: 0-89540-263-7.

THE KINGSHIP OF SELF-CONTROL by William George Jordan. The Kingship of Self-Control, The Crimes Of The Tongue, The Red Tape Of Duty, The Supreme Charity Of The World, Etc. 5x8. Pbk. ISBN: 0-89540-419-2.

THE MAJESTY OF CALMNESS by William George Jordan. The Majesty of Calmness, Hurry - the Scourge of America, The Power of Personal Influence, The Dignity of Self-Reliance, Failure as a Success, Doing Our Best at All Times, The Royal Road to Happiness. 63 pgs. 5x8. Pbk. ISBN: 0-98540-423-0.

A MESSAGE TO GARCIA and Other Essays by Elbert Hubbard. A Message to Garcia, The Boy from Missouri Valley, Help Yourself by Helping the House. "He was of big service to me in telling me the things I knew, but which I did not know I knew, until he told me." *Thomas A Edison* 48 pgs. 5x8. Pbk. ISBN: 0-89540-305-6.

POSITIVE THOUGHTS ATTRACT SUCCESS by Mary A. Dodson and Ella E. Dodson. "Unless We Can Do The Work Better, We Have No Right To Find Fault When Another Does It," "I Am a Holy Temple, and Send Out Love and Good To All The World," "What You Accomplish is Often

Determined by What You Attempt", "I will Develop a Powerful Personality," Etc. 64 pgs. 5x8. Pbk. ISBN: 0-89540-299-8.

THE SCIENCE OF GETTING RICH or Financial Success Through Thought by Wallace D. Wattles. The Right to be Rich, There is a Science to Getting Rich, How Riches Come to You, Thinking a Certain Way, How to Use the Will, Efficient Action, Getting into the Right Business, Etc. 160 pgs. 5x8. Pbk. ISBN: 0-89540-300-5.

THE SUCCESS PROCESS by Brown Landone. Five Factors Which Guarantee Success, The Process of Vivid Thinking, Tones Used in Persuading, Use of Action, Overcoming Hindrances, Developing Capacities, Securing Justice, Augmenting Your Success by Leadership, Etc. 233 pgs. 5 x 8. Hardback. ISBN: 0-89540-181-9.

VISUALIZATION AND CONCENTRATION by Fenwicke L. Holmes. The Creative Power of Mind, Metaphysics and Psychology, Mental Telepathy, Visualization and Dramatization, Concentration, How to Choose a Career. 152 pgs. 5x8. Pbk. ISBN: 0-89540-214-9.

THE HEART OF THE NEW THOUGHT by Ella Wheeler Wilcox. Let the Past Go, Thought Force, Opulence and Eternity, Morning Influences, Philosophy of Happiness, Common Sense, The Object of Life, Wisdom and Self Conquest, Concentration and Destiny, The Breath, Generosity and Balance, Etc. 92 pgs. 5x8. Pbk. ISBN: 0-89540-268-8.

YOU CAN BE WEALTHY TOO! by R.E.A. Lambert. "Wealth is nothing more nor less than common sense continuously applied to the use of small sums of money over a long period of time." Chapters include: The Secret of Wealth, The Six Foundation Stones, The Five Main Reasons for Saving, The Twelve Rules of Spending, The Seven Rules of Investment. 103 pgs. 5x8. Pbk. ISBN: 0-89540-459-1.

27 FEB 2007 B-5 Supp.doc